AC and 3-Phase

Simulations and Experiments

ZAP Studio, LLC

i

Cover Picture

Three-phase electric power lines along I-5 between Seattle and the US/Canada border.

Taken by Wing-Chi Poon. Licensed under the Creative Commons Attribution-Share Alike 2.5 Generic license.

AC and 3-Phase
Simulations and Experiments

by Sid Antoch

ZAP Studio, LLC · · · Philomath, Oregon

ISBN: 978-1-935422-14-3

Published by:

ZAP Studio, LLC
PO Box 1150
Philomath, OR 97370

www.zapstudio.com

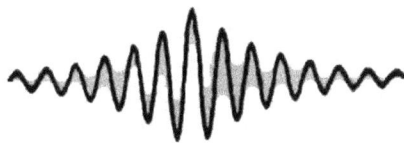

Contents

Resources

Parts
www.mouser.com
www.digikey.com
www.jameco.com
www.newark.com
www.alliedelec.com
www.allelectronics.com

Equipment
www.agilent.com
www.tek.com

USB Instruments
www.easysync-ltd.com
www.picotech.com

Simulation Software
Download free *LTspice* simulation software from
Linear Technology: www.linear.com

Introduction

This book may be used by any reader who wishes to learn by example and experiment. Simulation examples are presented which may be done using *LTspice*, a simulation program available as a free download from *Linear Technology.* Experiments provided may be performed using a solder-less breadboard, inexpensive parts, oscilloscope, function generator, and a low voltage 3-phase source.

All of the Three-phase experiments may be done with a 12 volt peak-to-peak, line to neutral, source capable of supplying up to 125mW per phase. This source may be easily built on a breadboard using the circuit provided in the appendix. This circuit may also be purchased assembled or as a kit from ZAP Studio, LLC: *www.zapstudio.com*.

All of the experiments demonstrate basic single-phase and three-phase principles. Analysis suggestions are provided at the end of each experiment.

The reader should be familiar with DC circuit analysis and have basic knowledge of AC circuits and phasor algebra. This book may be used as a supplement to an AC circuits course or for independent study.

Sid Antoch

Chapter 1: AC Circuit Basics

All components, resistors, capacitors, and inductors, will be assumed to be ideal in this book. Inductors used in the experiments will be considered to have an ideal inductance plus a series resistance.

In practice, all components have properties of resistance, capacitance, and inductance, although as is demonstrated by the experiments in this book, treating the components as ideal elements produces relatively accurate results at the low frequencies used. This is not always the case, especially at higher frequencies.

AC Circuit Elements

The basic properties of resistance, inductance, and capacitance are summarized here.

Resistance

Ohm's Law is the fundamental time domain equation for resistance. The voltage developed across a resistance is directly proportional to the current through it:

$$v(t) = i(t)\,R \quad \text{R is in ohms, t is in seconds.}$$

A steady-state sinusoidal time domain current may be expressed as:

$$i(t) = I\,sin(\omega t), \quad \text{where I is the peak current.}$$

A steady-state sinusoidal time domain voltage may be expressed as:

$$v(t) = i(t)\,R = IR\,sin(\omega t) = V\,sin(\omega t), \quad \text{where V is the peak voltage.}$$

The voltage across a resistor is in phase with the current through it. In the phasor domain the equation for resistance for steady-state AC voltages and currents is:

$$\mathbf{V} = \mathbf{I} \cdot R \quad \mathbf{V} \text{ and } \mathbf{I} \text{ are phasors.}$$

Capacitance

The fundamental time domain equation for capacitance is:

$$i(t) = C \frac{d\,v(t)}{dt}. \quad \text{C is in Farads and t is in seconds.}$$

Substitute $v(t) = V\sin(\omega t)$ into the previous equation:

$$i(t) = C\frac{d\,v(t)}{dt} = CV\frac{d}{dt}\sin(\omega t) = \omega CV\cos(\omega t).$$

The above result shows that the current through a capacitor leads the voltage across it by 90 degrees. Substitute $i(t) = I$, the peak current, when $t = 0$ in the above equation:

$$i(t) = \omega CV\cos(\omega t) \implies I = \omega CV \text{ at } t = 0.$$

$$\text{Define the capacitive reactance, } X_C = \frac{V}{I} = \frac{1}{\omega C}.$$

The fundamental phasor domain equation for capacitive reactance for steady-state AC voltages and currents is:

$$\mathbf{V} = \mathbf{I} \cdot \mathbf{X_C} \quad \mathbf{V}, \mathbf{I}, \text{ and } \mathbf{X_C} \text{ are phasors, } \mathbf{X_C} = -jX_C.$$

$\mathbf{V}$ is a phasor voltage and $\mathbf{I}$ is a phasor current. X_C, the capacitive reactance, is the capacitor's opposition to steady state AC current flow and is expressed in ohms.

Steady-state phasor analysis is used throughout this book. Single frequency and constant amplitude voltage sources are used for all experiments, simulations, and analysis.

Inductance

The fundamental time domain equation for inductance is:

$$v(t) = L\frac{d\,i(t)}{dt}. \quad \text{L is in Henries and t is in seconds.}$$

Substitute i(t) = $I sin(\omega t)$ into the previous equation:

$$v(t) = L\frac{di(t)}{dt} = CI\frac{d}{dt}sin(\omega t) = \omega CI cos(\omega t).$$

The above result shows that the voltage across an inductor leads the current through it by 90 degrees.

Substitute v(t) =V, the peak voltage, when t = 0 in the above equation:

$$v(t) = \omega L I cos(\omega t) \implies V = \omega LI \text{ at } t = 0.$$

$$\text{Define the inductive reactance, } X_L = \frac{V}{I} = \omega L.$$

The fundamental phasor domain equation for inductive reactance for steady-state AC voltages and currents is:

$$\mathbf{V} = \mathbf{I} \cdot \mathbf{X_L} \qquad \mathbf{V, I, \text{ and } X_L} \text{ are phasors, } \mathbf{X_L} = j\mathbf{X_L}.$$

V is a phasor voltage and **I** is a phasor current . **X_L**, the inductive reactance, is the inductor's opposition to steady-state AC current flow and is expressed in Ohms.

Phasors and Equations

This book is entirely about steady-state AC circuits so that phasor algebra will be adequate for circuit analysis. A scientific calculator such as the *Texas Instrument TI-89* or math program such as *Maple* may be used to solve phasor equations.

Inductive reactance will always be denoted as jX_L where $j = \sqrt{-1}$.

Capacitive reactance will always be denoted as $-jX_C$ where $-j = -\sqrt{-1}$.

Resistance will be denoted by its value in ohms.

Impedance **Z** will be designated by a rectangular or polar complex number, $R \pm jX$ or $|Z|\angle\theta$ where R is the magnitude of the resistance, X is the magnitude of the reactance, $|Z|$ is the magnitude of the impedance, and $\angle\theta$ is the phase angle of the impedance in degrees.

3

With phasors, AC circuit equations are written and solved by the same methods used for DC circuits. A basic knowledge of complex algebra is required to solve the equations.

RLC Example Circuit

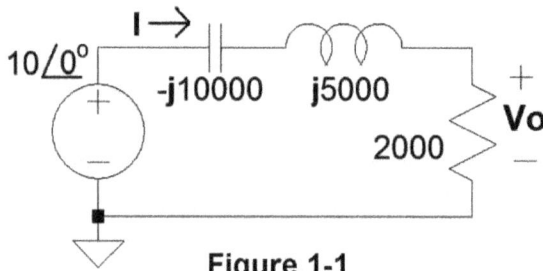

Figure 1-1

The voltage source in the RLC circuit in figure 1-1 supplies a 400Hz, 10V peak amplitude sine wave with a phase angle of 0 degrees. The circuit's output voltage, **Vo**, is calculated below.

First calculate the impedance, **Z**, of the series circuit:

$$Z = -j10000 + j5000 + 2000 = 2000 - j5000.$$

Next calculate the circuit current, **I**:

$$I = \frac{10}{2000 - j5000} = (0.6897 + j1.724)\text{mA (peak)}, \quad \text{(Rectangular format)}.$$

Finally calculate the voltage **Vo**:

$$Vo = IR = (0.6897 \times 10^{-3} + j1.724 \times 10^{-3})(2000) = (1.34 + j3.45)\text{volts (peak)}.$$

This result needs to be converted to polar form in order to compare it to actual circuit measurements. The peak amplitude and phase angle of **Vo** may be measured in the actual circuit with an oscilloscope. Peak amplitude is obtained using Pythagorean theorem and the phase angle is obtained using the tangent function as shown below.

$$|Vo| = \sqrt{1.34^2 + 3.45^2} = 3.7V_{peak} \qquad \theta = \arctan\left(\frac{3.45}{1.34}\right) = 68.8^0$$

4

The previous calculations may be performed on a calculator, like the *Texas Instruments TI-89*, or by a computer software program such as *Maple* from *Mathsoft*. The circuit may also be simulated using software such as *PSpice* or *LTspice*. Examples are presented below.

Solution using TI-89 Calculator

Set the calculator MODE: *Angle* to DEGREE, *Exponential Format* to ENGINEERING, *Complex Format* to POLAR. This is ideal for phasor calculations. Phasors can be entered in rectangular or polar format, but the output will always be polar, expressing the magnitude and phase angle of the result.

Enter the expression for the circuit current (The imaginary **i** must be used. It's the 2nd function of the CATALOG key):

10/(-**i**10000+**i**5000+2000) ENTER ►1.857E-3 ∠68.2

The result for the current can be multiplied by the resistance to get the output voltage.

ans(1) * 2000 ENTER ►3.714E0 ∠68.2

Circuit Simulation, Phasors, and LTspice

Simulation examples using LTspice and PSpice are presented throughout this book. LTspice is available as a free download from *Linear Technology*. It is a compact easy to use program and is especially recommended to those new to using simulation software.

   http://www.linear.com/designtools/software/

PSpice is available from *Cadence Design Systems, Inc.* It is a large and expensive program, but a free evaluation version is available.

   http://www.cadence.com/products/orcad/pages/downloads.aspx

It is not the intention of this book to teach how to use simulation software. Instructions and tutorials on the use of most software programs are readily available as part of the software package.

RLC Example Circuit Simulation

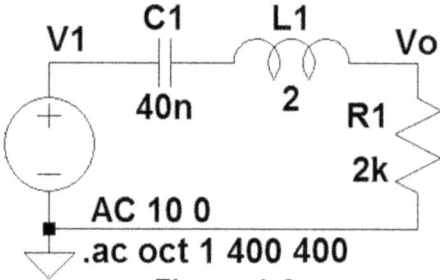

Figure 1-2

The circuit in Figure 1-2 approximates the "RLC Example Circuit" that we just analyzed. However, instead of specifying the reactances of the capacitor and inductor, we need to specify a part value. A 40nF capacitor and a 2H inductor are used in this circuit. The analysis is at 400Hz, so the reactances of the capacitor and inductor are given by:

$$X_C = \frac{1}{2\pi 400(40\times 10^{-9})} = 9947\Omega \quad X_L = 2\pi 400(2) = 5027\Omega$$

Source V1 is set to AC with an amplitude of 10, which represents volts peak in this simulation. The phase angle of the source is set to 0 degrees.

The "Edit Simulation command" dialog box shown on the right is accessed from the main menu under "Simulate".

Analysis type is set to AC at a frequency of 400Hz.

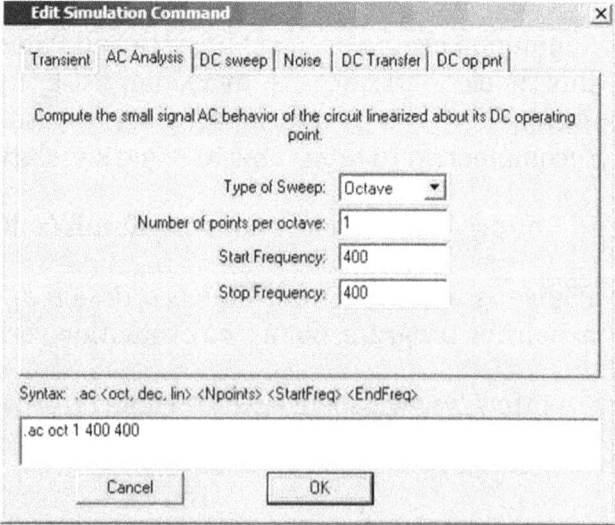

6

Results for this analysis should be about the same as the previous analysis results of the example circuit.

The "Net Labeling" tool was used to identify the output voltage node: **Vo**. Net labels make it easier to identify the nodes in the analysis results. Otherwise LTspice will assign its own node labels, such as n(001), n(002), etcetera. The node locations would then have to be determined from the circuit's "*netlist*".

The *netlist* is part of the "command line" language of Spice. It is available from the main menu under "*View*". Select "*SPICE Netlist*" to view it.

The simulation results below are in close agreement with the calculated results for the example circuit.

```
            --- AC Analysis ---

    frequency:      400                 Hz
    V(n001):      mag:          10 phase:           0°
    V(vo):        mag:     3.76537 phase:      67.8806°
    V(n002):      mag:      10.185 phase:      136.184°
    I(C1):        mag: 0.00188269 phase:      67.8806°
    I(L1):        mag: 0.00188269 phase:      67.8806°
    I(R1):        mag: 0.00188269 phase:      67.8806°
    I(V1):        mag: 0.00188269 phase:     -112.119°
```

Note: It is possible that the phase angles of the currents may be given exactly 180 degrees out of phase of what they should be. This is because the current direction arrow is given with respect to the part's terminals, always from plus to minus.

However, resistor, capacitor, and inductor terminals are not labeled on the schematic. Bottom line: If the simulated current is 180 degrees out of phase with the actual current for a part, that part can be turned around on the schematic (two 90 degree rotations).

Voltage phase angles are not a problem since voltages are always given with respect to ground. Currents can be calculated from the voltages, especially if the current direction is uncertain from the simulation.

Working with Phasors

A sinusoidal voltage has a value that is a function of time and may be expressed as:

$$v = V \sin(\omega t + \theta) \quad \text{or} \quad v = V \cos(\omega t + \theta) .$$

V is the peak value of the voltage. The instantaneous value of the voltage, v, is a function of time t, and the constant angle, θ. The angle θ is usually expressed in degrees; however, the angle "ωt" has units of radians. To evaluate the argument of the sine or cosine, first convert angle "ωt" to degrees and then add the angle "θ".

Figure 1-3 below shows phasor V_a rotating counter clockwise at angular velocity ω. The value of V_a at time t is its vertical component at time t, and is equal to v_a.

Figure 1-3 shows that v_a equals 0 when t is zero. It reaches a maximum value when t equals 0.25mSec.

V_b has a smaller peak value than V_a and is shown lagging V_a by about 50 degrees. V_b is expressed as a negative angle because it lags V_a in time, in this example as -50 degrees.

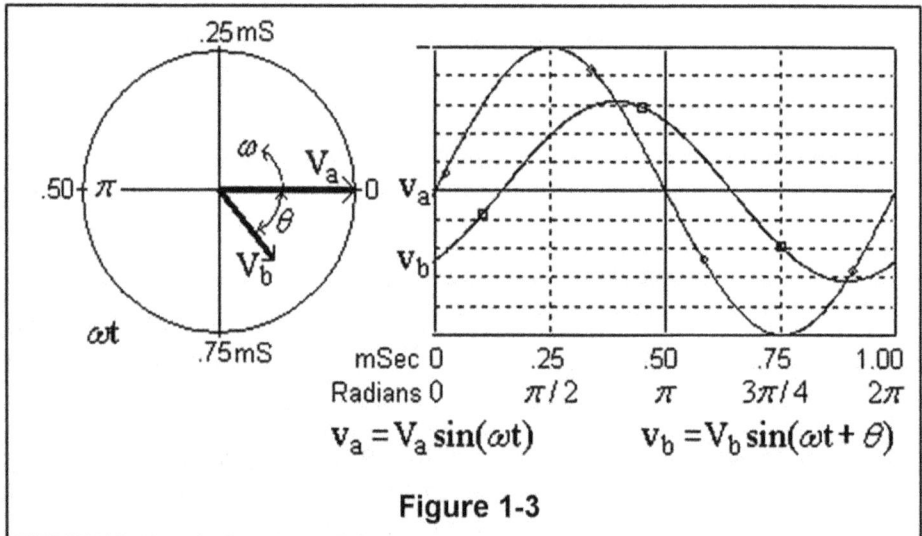

Figure 1-3

Frequency, f, of V_a and V_b is 1KHz (ω=6283 r/Sec). Period, T, is 1.0mS.

Given that the magnitude of V_a is 12 volts peak and V_b is 7 volts peak, V_a and V_b can be expressed as phasors as:

$$V_a = |V_a| \angle \theta_a = 12 \angle 0^0 \quad \text{and} \quad V_b = |V_b| \angle \theta_b = 7 \angle -50^0$$

Figure 1-4 below also shows phasor V_a rotating counter clockwise at an angular velocity ω. The value of V_a at time t is its vertical component at time t, which is a cosine function of t and is equal to v_a. Figure 1-4 shows that v_a is maximum when t is zero, and zero when t equals 0.25mSec.

The only difference between Figure 1-3 and Figure 1-4 is the reference time t = 0. In figure 2 the t = 0 reference time is chosen when the value of V_a is maximum. Otherwise the phasors in Figure 1-3 and Figure 1-4 are identical.

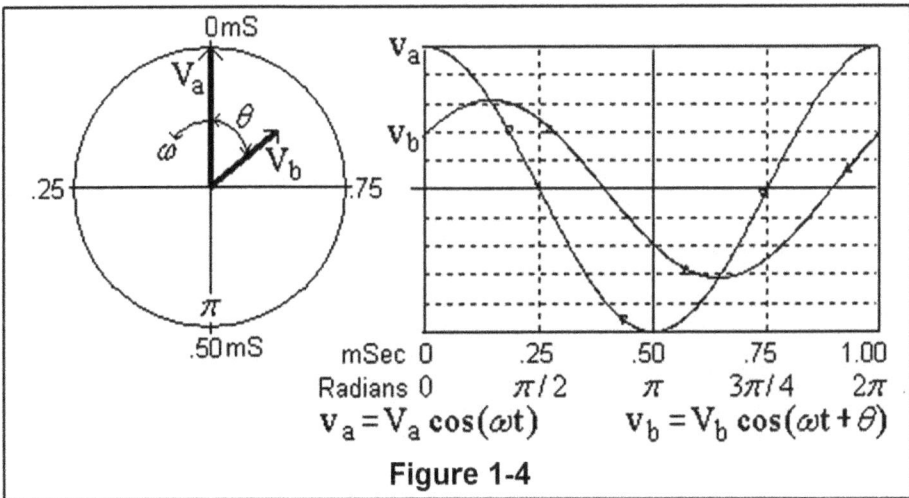

Figure 1-4

Some textbooks use the sine function as reference to describe sinusoids and others use the cosine function. Some books express the magnitude of a phasor in RMS units rather than peak units.

Magnitudes and Angles in LTspice

In LTspice, the magnitude of a sinusoidal source is expressed in peak units. For example, if you set the AC amplitude to 5 volts, a transient analysis graph will display the magnitude of the sinusoid as 5 volts peak or 10 volts peak-to-peak.

You can set the phase angle of the source by right clicking on it. This opens the property editor where you can specify a "*SINE*" source with the Phase angle value in degrees, "*Phi(deg)*".

Refer to figure 1-5 where the source amplitude is set to 5V. The first cycle of the waveforms at nodes n001 and n002 differ from the second cycle. It usually takes a few cycles for the waveforms to reach steady state (where each cycle is the same).

The phase angle of V1 was set to 0 degrees in the top simulation below and to 90 degrees in the bottom simulation. Refer to the command lines below the schematics: SINE(0 5 1k 0 0 0 3) and SINE(0 5 1k 0 0 90 3).

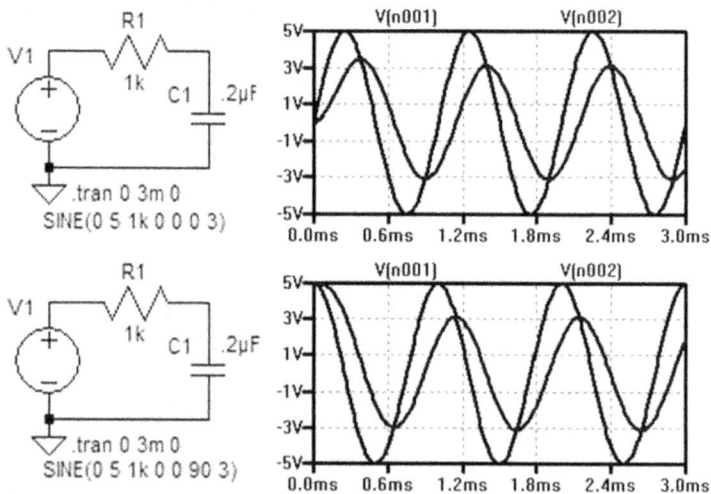

Figure 1-5

In this book "*AC Analysis*" will be the only analysis type used. *AC Analysis* is used for steady-state AC because it outputs the magnitudes and phase angles of the circuit voltages and currents. But it does not provide a graphical time-domain display.

It may be desirable to do a *Transient* analysis to compare the simulation results to the oscilloscope display. Right click on the voltage source in the schematic. This opens a dialog box where the properties of the voltage source can be specified.

Voltage source settings shown below are for the simulation in figure 1-5.

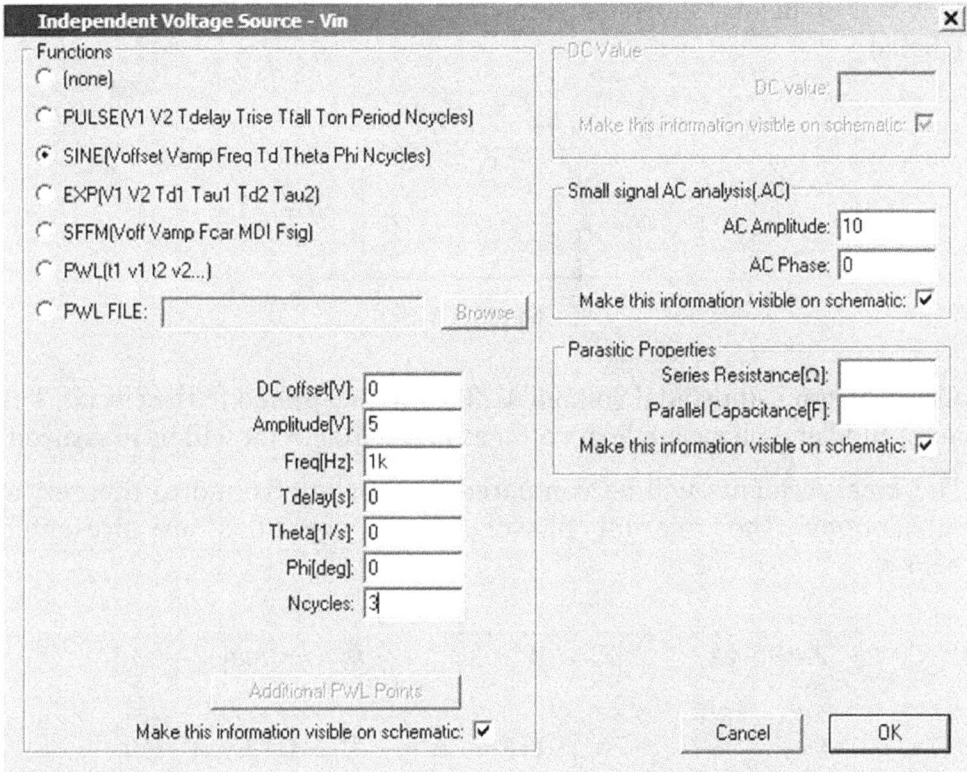

Suggested Exercises

1. Change the voltage source frequency in the RLC Example Circuit in Figure 1-2 to 200Hz and calculate the circuit's current and the voltages across the circuit elements.

2. Simulate the RLC Example Circuit in Figure 1-2 at 200Hz. Simulate a homework problem from your textbook involving.

3. In Figure 1-2, change the value of the capacitor so that the magnitude of its reactance is equal to the magnitude of the reactance of the inductor (at 400Hz). Calculate the resulting current and the voltage across R1.

4. Verify your calculations in exercise 3 above by simulating the circuit.

Experiment 1a: Series RC Circuit Measurements

Figure 1-6 below shows a series RC circuit with circuit properties labeled.

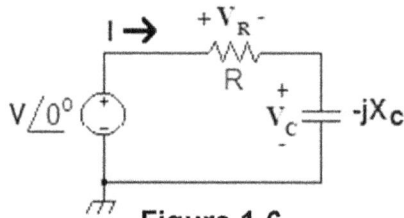

Figure 1-6

A steady-state sinusoidal voltage, $V\angle0°$, will be applied to the circuit. The amplitude and phase angle of voltages in the RC circuit will be measured.

The measurements will be compared to simulations and to theoretical calculations. The relevant phasor-domain equations are presented below.

$$Z = R - jX_C \qquad |Z| = \sqrt{R^2 + X_C^2} \qquad \theta_Z = arctan\left(-\frac{X_C}{R}\right)$$

$$I = \frac{|V|\angle0°}{Z} \qquad V_R = I \cdot R \qquad V_C = I(-jX_C)$$

Equipment and Parts

Function Generator, Oscilloscope, and Breadboard.
C = 100nF, 5%. R = 1000Ω, ¼ watt, 5%.

For greater accuracy, measure the values of R and C.

R = _____   C = _____

Resistance values can be measured accurately with most digital multi-meters. Use a 5% or better tolerance capacitor if you can't measure the capacitance.

If you do not measure part values, an error analysis may be done on the results using the tolerance values of the parts.

Procedure

1. Connect the circuit in Figure 1-7a below. Connect oscilloscope channel 1 to measure V_S and channel 2 to measure V_C. Set the generator to produce a 1000Hz, 6V p-p sine wave with no offset.

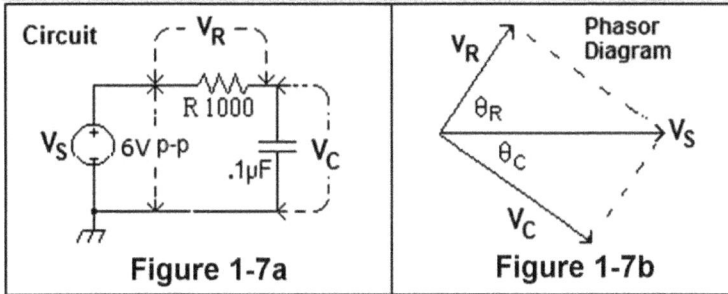

Figure 1-7a

Figure 1-7b

2. Set channels 1 and 2 to AC coupling and 1V/DIV. Set the horizontal to 100µS/DIV. Trigger on channel 1. Adjust the oscilloscope to get a display similar to Figure 1-8a below. V_S starts at 0V with a positive slope at the left side of the screen.

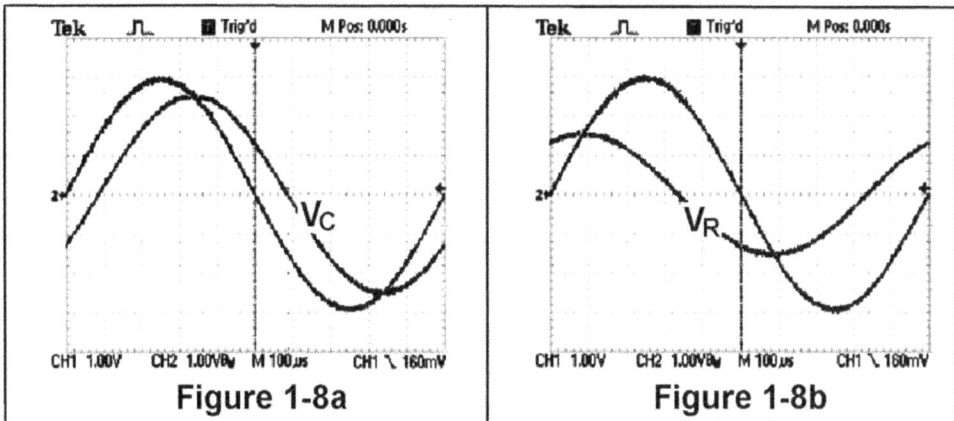

Figure 1-8a

Figure 1-8b

In Figure 1-8a the period of the sinusoid is exactly 1000µS. V_C crosses zero 89µS after V_S. Therefore V_C lags V_S. The angle can be calculated by the proportion below:

$$\theta_C = \left(\frac{-89\,\mu S}{1000\,\mu S} \right) 360^0 = -32.1^0.$$

13

In Figure 1-8b V_R crosses zero 160μS before V_S. Therefore V_R leads V_S. The angle can be calculated by the proportion:

$$\frac{160\,\mu S}{1000\,\mu S} = \frac{\theta_R}{360^0} \quad \text{or} \quad \theta_R = \left(\frac{160\,\mu S}{1000\,\mu S}\right)360^0 = 57.6^0.$$

The oscilloscope display in Figure 1-8b was obtained by swapping the positions of the resistor and capacitor so that one end of the resistor was connected to circuit ground. However, this does not need to be done since V_R can also be obtained by calculating the difference between V_S and V_C: $V_R = V_S - V_C$,

3. Measure and record the magnitude and phase angle of V_C.

    V_C _____volts p-p $\theta_C =$ _____degrees

4. Measure the magnitude and phase angle of V_R in your circuit by swapping the positions of the capacitor and resistor, or, calculate V_R using your results for V_C from step 3 above. Record below:

    V_R _____volts p-p $\theta_R =$ _____degrees

Analysis

Note: If possible, use the measured values of the components used in Experiment 1a to do the calculations and simulations below.

1. Calculate the theoretical value (magnitude and phase angle) of the current, I, in the series RC circuit.

2. Calculate the percent difference between the measured and calculated value of I (use Ohm's Law and the voltage measured across R to calculate the measured value of I).

3. Calculate the theoretical value (magnitude and phase angle) of the voltages, V_R and V_C, in the series RC circuit.

4. Calculate the percent difference between the measured and calculated value of V_R and V_C.

5. Simulate the series RC circuit to obtain the magnitude and phase angle of V_R and V_C.

LTspice Simulation Example: Series RC Circuit

Connect the circuit as shown in Figure 1-9. Label the nodes (N1, N2, and N0) to make it easier to interpret the results. Otherwise LTspice will number the nodes. The node labels will appear in the simulation results.

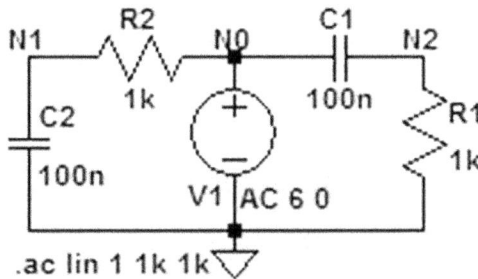

Figure 1-9

Click on "Simulate" in the main menu bar. Select "Edit Simulation Command", to open the dialog box shown in Figure1-10 below.

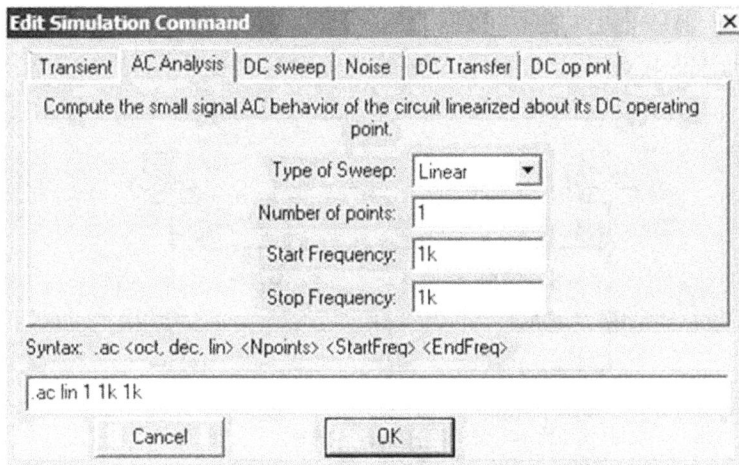

Figure 1-10

Select *AC Analysis* and enter values as shown. Note that this is for the one frequency of 1000Hz. You can specify the simulation to be done at more than one frequency, if desired. Run the simulation. A window will open listing the simulation results.

Simulation results

```
         --- AC Analysis ---

frequency:     1000          Hz
V(n2):         mag:    3.19211 phase:      57.8581°          voltage
V(n0):         mag:          6 phase:  -4.24074e-015°        voltage
V(n1):         mag:     5.0804 phase:     -32.1419°          voltage
```

PSPice Simulation Example: Series RC Circuit

This simulation was done with with an evaluation version of OrCAD PSpice. It's the same program as the commercial version but it has limiteds on the the number of circuit nodes and number and type of parts. Check with *Cadance* for more information.

http://www.cadence.com/products/orcad/pages/downloads.aspx

The PSpice circuit in figure 1-11 simulates simultaneously the original circuit and the circuit with the resistor and capacitor positions swapped.

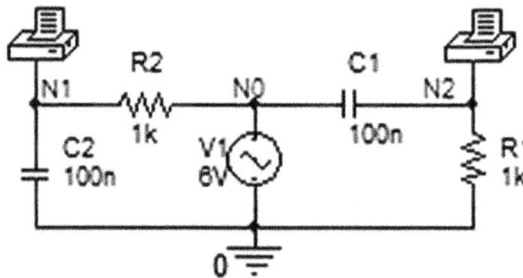

Figure 1-11

Steady-state analysis is used to obtain V_C and V_R.

V1 is a steady-state source, VAC, from the "Source" library. It was set to 6V which corresponds to the 6V p-p used in the experiment. The voltages in the simulation results will have peak-to-peak units.

16

Printers, "VPRINT1", are in the "SPECIAL" library. Printers in the schematic read the voltages with respect to ground, and list their values in the "Output File". Phase angles are with respect to the source, V1.

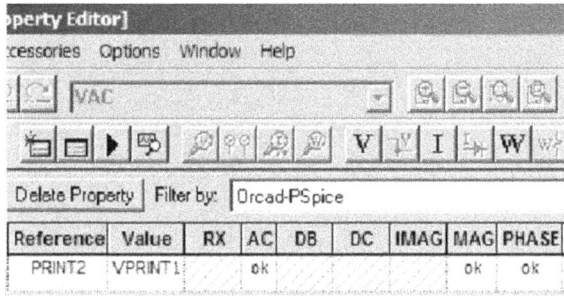

Figure 1-12

Double click on each printer to open its property editor. Set the following in the "Property Editor" as shown in Figure 1-12 above.

AC = ok, *MAG* = ok, and *PHASE* = ok

Select *Analysis Type*: AC Sweep/Noise.   *AC Sweep Type*: Linear.

Start frequency: 1kHz.   *Stop frequency*: 1kHz.   *Total Points*: 1.

View the "Output File" to see the results. See edited results below:

```
FREQ          VM(N2)        VP(N2)
1.000E+03     3.192E+00     5.786E+01

FREQ          VM(N1)        VP(N1)
1.000E+03     5.080E+00     -3.214E+01
```

Summary:

The simulated resistor voltage 3.19V p-p at an angle of 57.86 degrees.

The simulated capacitor voltage 5.08V p-p at an angle of -32.14 degrees.

Experiment 1b: Series-Parallel Circuit Measurements

Figure 1-13 below shows a series-parallel circuit with its circuit properties labeled.

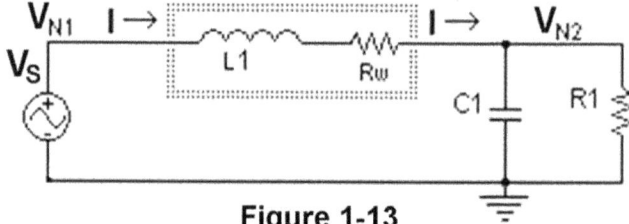

Figure 1-13

A steady-state sinusoidal voltage, **Vs**, will be applied to the circuit. The amplitude and phase angle of voltages in the circuit will be measured. The measurements will be compared to simulations and theoretical calculations.

The relevant phasor-domain equations are presented below.

$$Z = Rw + j\omega L1 + Z_P \qquad Z_P = \frac{-j\left(\dfrac{1}{\omega C1}\right)R1}{R1 - j\left(\dfrac{1}{\omega C1}\right)} \qquad I = \frac{|V_s|\angle 0^0}{Z} \qquad V_{N2} = I \cdot Z_P$$

$$\text{Node Voltage Method}: \quad \frac{V_{N2} - |V_s|\angle 0^0}{Rw + j\omega L1} + \frac{V_{N2}}{R1} + \frac{V_{N2}}{-j\left(\dfrac{1}{\omega C1}\right)} = 0$$

Equipment and Parts

Function Generator, Oscilloscope, and Breadboard.
C1 = 100nF, 5%. R1 = 4700Ω, ¼ watt, 5%. L1 = 100mH, 5%.
Measure the resistance of the inductor, Rw.

Rw = _____

For greater accuracy, measure the values of R1, C1, and inductance of L1

R1 = _____   C1 = _____   L1.= _____

Procedure

1. Connect the circuit in Figure 1-14.

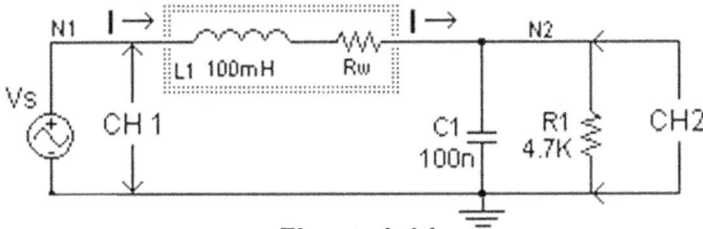

Figure 1-14

2. Connect channel 1 of the oscilloscope to node N1 and channel 2 to N2. Set the trigger to channel 1.

3. Set the function generator to produce a 3.0V p-p, 1600Hz, sine wave and adjust the oscilloscope to accurately measure the waveform amplitudes and phase angles at nodes N1 and N2.

   Measure and record the magnitude and phase angle of the voltage at node N2

   V_{N2} _____volts p-p θ_{N2} = _____degrees

Analysis

1. Calculate the theoretical value (magnitude and phase angle) of the voltage at node N2.

2. Calculate the theoretical value of the current **I**.

3. Calculate the approximate value of the current, **I**, using Ohm's Law, the voltage measured at node N2, and the impedance of R1 and C1 in parallel.

4. Calculate the approximate value of the current, **I**, using Ohm's Law, the voltage measured at node N2, and the impedance of L1.

5. Simulate the circuit and compare the results to your measurements and calculations.

LTspice Simulation Example: Series-Parallel Circuit

Connect the circuit shown in Figure 1-15.

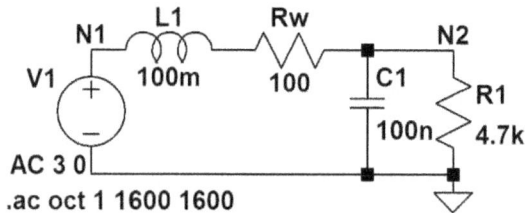

Figure 1-15

Right click on V1 to set its AC value to 3 volts and angle to 0 degrees. Set the AC analysis to 1600Hz.

AC analysis results:

```
        --- AC Analysis ---

frequency: 1600Hz
V(n1):     mag:          3    phase:   2.1e-015°  voltage
V(n001):   mag:    9.78565    phase: -82.4°       voltage
V(n2):     mag:     9.5357    phase: -88.0°       voltage
I(C1):     mag: 0.00958633    phase:   1.9°       device_current
I(L1):     mag: 0.00979868    phase: -10.0°       device_current
I(R1):     mag: 0.00202887    phase: -88.0°       device_current
I(Rw):     mag: 0.00979868    phase: -10.0°       device_current
```

PSpice Simulation Example: Series-Parallel Circuit

Connect the circuit as shown in Figure 1-16 below. The printers, "IPRINT" and "VPRINT1", are in the "SPECIAL" library.

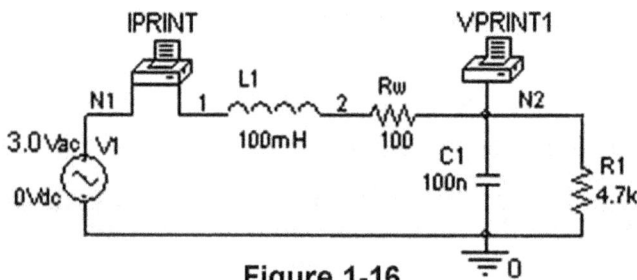

Figure 1-16

Enable the printers by double clicking on each printer to open the property editor. Type "ok" under AC, MAG, and PHASE.

Select *Analysis Type*: AC Sweep/Noise.   *AC Sweep Type*: Linear.

Start frequency: 1.6kHz. *Stop frequency*: 1.6kHz. *Total Points*: 1.

Results in Output File:

FREQ	IM(V_PRINT2)	IP(V_PRINT2)
1.600E+03	9.800E-03	-1.001E+01

FREQ	VM(N2)	VP(N2)
1.600E+03	9.536E+00	-8.806E+01

TI-89 example at 1600 Hz:

X_{L1} reactance in ohms:

2*π*1600*.1 `ENTER` ▶ 1005

X_{C1} reactance in ohms:

1/(2*π*1600*.1E-6) `ENTER` ▶ 995

Impedance **Z** in ohms:

100+1005i+(-4700*995i)/(4700-995i) `ENTER` ▶ (306.2∠9.91)

Chapter 2: AC Power and Power Factor

The power supplied to a reactive circuit, as measured by the product of the voltage and current applied to it, is not equal to the power that the circuit actually dissipates. This is because the average power dissipated by the reactive components is zero. However, the reactive component does contribute to the current supplied to the circuit.

The product of the circuit current and voltage is called "apparent power" and is designated by the phasor **S**. Its magnitude is measured in Volt-Amps. **S** is a complex number where in rectangular coordinates the real part is the circuit's actual dissipated power and the imaginary part is the power associated with the circuit's reactance.

Reactive power is designated by the phasor **jQ**. Its phase angle is plus 90 degrees if it is inductive and minus 90 degrees if it is capacitive. Its magnitude is measured in Volt-Amps Reactive, or VARs for short.

Capacitive Circuit

Figure 2-1a below shows a series RC circuit with the voltages and current labeled. The circuit current **I** is set to 0 degrees so that the phase angle of the resistor voltage and capacitor voltage will be relative to a current at 0 degrees.

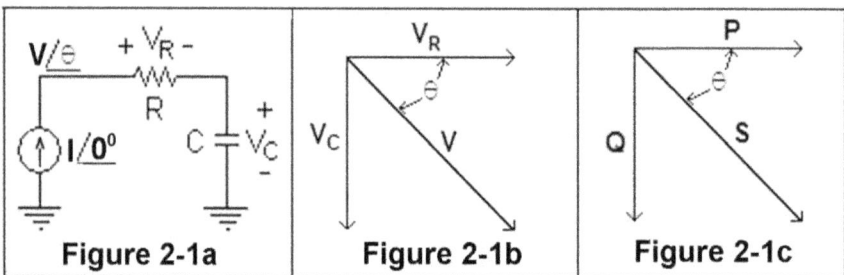

Figure 2-1a   Figure 2-1b   Figure 2-1c

Figure 2-1b shows a voltage phasor diagram for the series RC circuit. The resistor voltage is in phase with the circuit current and the capacitor voltage lags the resistor voltage and current by 90^0.

Figure 2-1c shows the phasor diagram for the circuit power. **Q** is at -90 degrees and **S** is at a negative angle, θ. This circuit is said to have a leading power factor because the current leads the voltage.

Figure 2-1: Relevant Equations

Applying trigonometry to figure 2-1 yields the following set of equations.

$$V_R = |V|\cos\theta \quad V_C = j|V|\sin\theta \quad P = |S|\cos\theta \quad Q = j|S|\sin\theta$$

The sign of the "**j**" terms is determined by the sign of the phase angles. **Q** is positive for an inductive reactance and negative for a capacitive reactance.

Note that the magnitudes of the voltages and currents must be in RMS units when calculating the apparent power **S**, reactive power **Q**, and average power, P.

Inductive Circuit

Figure 2-2a below shows a series RL circuit with the voltages and the current labeled. Again the angle of the current is set to 0 degrees so that the angle of the resistor voltage and capacitor voltage will be relative to a current of 0 degrees.

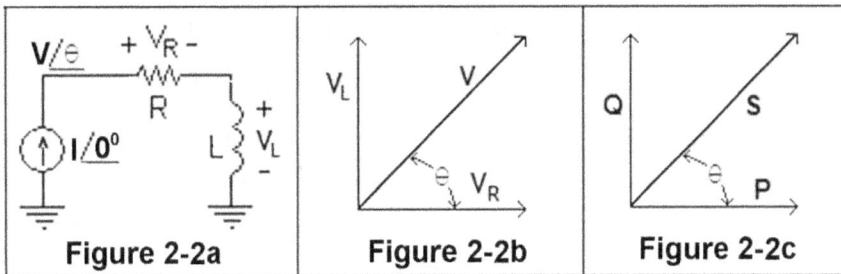

| Figure 2-2a | Figure 2-2b | Figure 2-2c |

Figure 2-2b shows a voltage phasor diagram for the series RL circuit. Resistor voltage is in phase with the current. The inductor voltage leads the resistor voltage and the current by 90⁰.

Figure 2-2c shows the power phasor diagram for the circuit. This circuit has a lagging power factor because the current lags the voltage.

The angles for the component powers correspond to the angles for the component voltages. For this correspondence to be true, the apparent power **S** is calculated as the product of the applied voltage times the complex conjugate of the circuit current: **S = V I***.

Figure 2-2: Relevant Equations

Applying trigonometry to figure 2-1 yields the following equations.

$$\mathbf{V_R} = |\mathbf{V}|\cos\theta \quad \mathbf{V_L} = \mathbf{j}|\mathbf{V}|\sin\theta \quad \mathbf{P} = |\mathbf{S}|\cos\theta \quad \mathbf{Q} = \mathbf{j}|\mathbf{S}|\sin\theta$$

The sign of the "**j**" terms is determined by the sign of the phase angles.

Note that the magnitudes of the voltages and currents must be in RMS units when calculating the apparent power **S**, reactive power **Q**, and average power, P.

Power Factor

The power factor of an electric circuit is the ratio of the circuit's average power P to the magnitude of its apparent power **S**.

$$pf = \frac{P}{|\mathbf{S}|} \quad \text{also} \quad pf = cos(\theta)$$

θ is the angle between the circuit's voltage and current.

Power factor is a dimensionless number between zero and one.

An inductive load or circuit is said to have a lagging power factor because the circuit current lags the circuit voltage.

A capacitive load is said to have a leading power factor because the circuit current leads the circuit voltage.

In an electric power system, a load with a high power factor is desirable because a low power factor load requires more current to transfer the same amount of useful power than a high power factor load.

Power factor compensation is often applied to low power factor loads and power distribution systems to reduce transmission and distribution line currents and increase power transfer efficiency.

Experiment 2a: AC Power

The voltage magnitudes and phase angles across the resistor and capacitor in a series RC circuit will be measured. Resistor and capacitor powers, P and **Q**, will be calculated from these measurements. The power relationship in the series RC circuit will be investigated and compared to theoretical calculations.

Equipment and Parts

Function Generator, Oscilloscope, DMM, and Breadboard.
C = 100nF, 5%. R = 4700Ω, ¼ watt, 5%.

For greater accuracy, measure the values of R1, and C1.

R1 = _____   C1 = _____

Procedure

1. Connect the circuit in Figure 2-3a. **Vs** is the function generator. Connect oscilloscope channel 1 to node n1 and channel 2 to node n2.

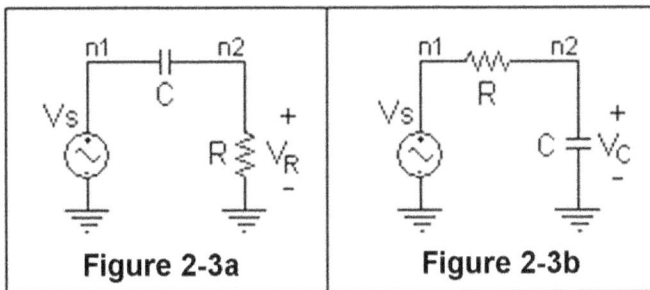

Figure 2-3a   Figure 2-3b

2. Set the function generator to produce a 6V p-p, 1000Hz sine wave with no offset. Set the oscilloscope to observe nodes n1 and n2 so that the magnitudes and phase angles of the voltages can be measured accurately.

3. Measure and record the peak-to-peak voltage and phase angle of V_R at node n2.

V_R mag = _____V p-p.

V_R Phase = _____degrees. Leading or lagging?

4. Set the DMM to measure AC volts. Measure and record the voltage **Vs** at node n1 and **V_R** at node n2 with the DMM.

DMM Measurements: **Vs** = _____Vrms.   **V_R** = _____Vrms.

Check that the measurement of the magnitudes of **Vs** and **V_R** agree with the oscilloscope measurements of **Vs** and **V_R**.

$$V_{S(p\text{-}p)} = 2\sqrt{2}\,V_{S(RMS)} \quad \text{and} \quad V_{R(p\text{-}p)} = 2\sqrt{2}\,V_{R(RMS)}$$

5. Connect the circuit in Figure 2-3b. Simply swap the positions of the resistor and capacitor, everything else remains the same.

   Measure and record the peak-to-peak voltage **Vc** at node n2.

   Vc = _____volts peak-to-peak.

6. Measure and record the phase angle of the voltage **Vc** at node n2.

   Vc Phase = _____degrees. Leading or lagging?

7. Set the DMM to measure AC volts. Measure and record the voltage Vc at node n2 with the DMM.

   DMM Measurement: **Vc** = _____Vrms.

Analysis

Note that the purpose of the two circuits was to be able to measure the resistor voltage   and the capacitor voltage with respect to the common grounds of the function generator and oscilloscope.

Both circuits have the same impedance and therefore the same current. The current   can be calculated using the voltage measured across the resistor $(I = V_R/R)$.

1. Calculate and record peak-to-peak magnitude of **I** in mA and the phase angle of **I** in degrees using the voltage and phase angle of **V_R** measured in step 3 of the procedure. This is the measured current because it is calculated from the measured voltage.

   I = _____mA p-p.   **I** Phase = _____ degrees.

2. Convert the magnitude of **I** to RMS units: **I** = _____mA RMS.

3. Calculate and record the apparent power, **S**, supplied to the circuit by the source, **Vs**. Use the measured current from analysis step 2. Express the result in mVA and in polar form (magnitude and phase angle). Use RMS units for all voltages and currents. Use the phase angle measured in procedure step 1.

    |**S**| = _____mVA (Measured apparent power).

4. Calculate and record the average power, P, and the reactive power, **Qc**, by expressing **S** in analysis step 3 above in rectangular form.

    P = _____mW   **Qc** = _____mVAR.

5. Calculate P and the magnitude of **Qc** using values measured by the DMM in procedure steps 4 and 7. Compare the results to the analysis step 4 result.

6. Calculate the theoretical value of the apparent power, **S**. Compare the result to the oscilloscope measurement result of analysis step 3 above.

7. Express the percent difference between the theoretical and measured results for the power dissipated by the resistor and for the reactive power of the capacitor.

8. Simulate the circuit and compare the results to your calculations and measurements.

9. Use a spreadsheet to compare the results of the measurements, calculations, and simulations.

LTspice Example

This example uses different part values than the experiment. Use this as a guide to simulate your experiment.

Figure 2-4 shows the LTspice circuit diagram.V1 is a 4Vrms AC source whose frequency is set to 1200Hz.

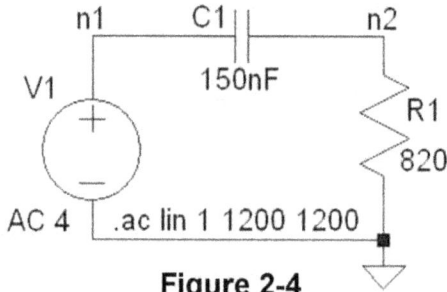

Figure 2-4

The AC analysis results are given below. The voltage at node n2 leads the voltage at node n1 by 47 degrees. The current through R1 (and the circuit) is in phase with the voltage at node n2.

```
    --- AC Analysis ---

frequency: 1200Hz
V(n2):      mag:      2.71996 phase: 47.1572°
V(n1):      mag:            4 phase:       0°
I(R1):      mag: 0.00331702 phase: 47.1572°
```

The LTspice analysis results are used to calculate **S**, P, and **Q**.

$$I = .003317\angle 47.16^0 \quad \text{therefore} \quad \boldsymbol{I}^* = .003317\angle -47.16^0$$

$$\boldsymbol{S} = \boldsymbol{Vs}\boldsymbol{I}^* = 4(.003317\angle -47.16^0) = 13.27mVA = 9.02 - j9.73\,mVA$$

P=9.02mW dissipated. Q=9.73mVARs capacitive.

Experiment 2b: Series Compensation

Electrical power is transmitted and distributed by long wires and transmission lines. The wire's resistance is proportional to its length, but the power loss in the wire is proportional to the square of its current. Long wires also have a significant inductive and capacitive reactance. The voltage developed across the inductance of the wire reduces the voltage and power supplied to the load.

Series compensation is used to cancel the effect of wire inductance. Capacitance in series with an inductive transmission can be used to increase the power delivered to the load. The reactance of the capacitance must be approximately equal to the inductance of the transmission line. This experiment also demonstrates the effect of under compensation and over compensation on the power delivered to the load.

Equipment and Parts

Function Generator, Oscilloscope, and Breadboard
R2: 560Ω, ¼ watt, 5%. Cx: Three 0.1uF, 5%.
L: 56mH, 5%. Mouser part #434-02-563J (see appendix 2).

Use 5% tolerance parts. The part values do not need to be measured.

Procedure

1. Connect the circuit in figure 2-5 below with Cx = .1uF. Connect a jumper wire across the capacitor, Cx. This will be the uncompensated circuit.

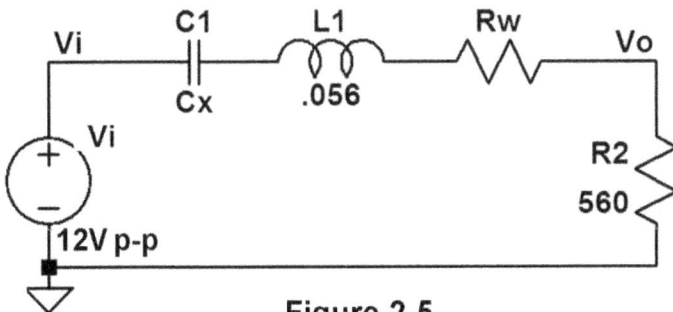

Figure 2-5

3. Set the function generator to produce a 12.0V peak-to-peak, 1500Hz, sine wave as measured by the oscilloscope channel 1.

4. Set up a spreadsheet for your measurements as shown below:

	A	B	C	D	E	F	G	H	I	J
1	Circuit	t mS	Vo mag	Vo θ	I mA	I θ	Si mVA	Pi mW	Po mW	% Eff.
3	Comp 0									
4	Comp 1									
5	Comp 2									
6	Comp 3									

Note: Columns D through J will be calculated by the spreadsheet. Instructions for the calculations are given in the analysis section of this lab exercise.

5. Connect channel 2 of the oscilloscope to measure Vo. Measure the peak-to-peak value of the voltage Vo, convert it to RMS, and record in cell C3.

   Measure the time between the positive slope zero crossings of Vi and Vo (in milliseconds). Record the result in cell B3. The time is negative if Vo crosses zero after Vi.

6. Remove the jumper wire across the .1uF capacitor. This will be the compensated circuit "Comp 1" with Cx = .1uF.

7. Repeat step 5 but enter the measurements in cells C4 and B4.

8. Connect a .1uF capacitor in parallel with the .1uF capacitor on the board so Cx = .2uF.

9. Repeat step 5 but enter the measurements in cells C5 and B5.

10. Connect another .1uF capacitor in parallel with the two .1uF capacitors on the board so Cx = .3uF.

11. Repeat step 5 but enter the measurements in cells C6 and B6.

Analysis

1. Enter the following equations into the indicated cells:

   Cell D3: =(B3/667)*360   Cell E3: = C3/.560   Cell F3: =D3

   Cell G3: = 4.242*E3   Cell H3: = G3*cos(3.14*F3/180)

   Cell I3: =C3*E3   Cell J3: =(I3/H3)*100

   Verify that the above equations are correct. Note that **Vo** is in volts peak-to-peak. **I** is in milliamps RMS.

2. The example spreadsheet below was produced with R2 = 510 ohms using simulation data and the labeled values of capacitance and inductance. Your results should be similar.

	A	B	C	D	E	F	G	H	I	J
1	Circuit	t µS	|Vo|	Vo θ	|I mA|	I θ	Si mVA	Pi mW	Po mW	% Eff.
3	Short	-74.80	2.93	-40.39	5.75	-40.39	24.36	18.56	16.83	90.70
4	Cx=.1	79.80	2.77	43.09	5.43	43.09	23.03	16.82	15.04	89.43
5	Cx=.2	0.00	3.79	0.00	7.43	0.00	31.51	31.51	28.16	89.39
6	Cx=.3	-31.40	3.63	-16.96	7.12	-16.96	30.18	28.87	25.84	89.50

3. Simulate the circuit for each value of Cx using AC analysis. Set the sweep to octave, 1 point per octave, start frequency to 1500Hz and stop frequency to 1500Hz. Refer to the example simulation in the next section.

4. Explain the results for Vo in terms of the circuit's power factor.

   Calculate the percent difference between the simulated results and the measured results for **Vo**.

5. Given the ±5% tolerance of the inductor and capacitor values, calculate the possible range of the circuit's impedance..

LTspice Example

This example uses different value of R2 than the experiment. Use this as a guide to simulate your experiment.

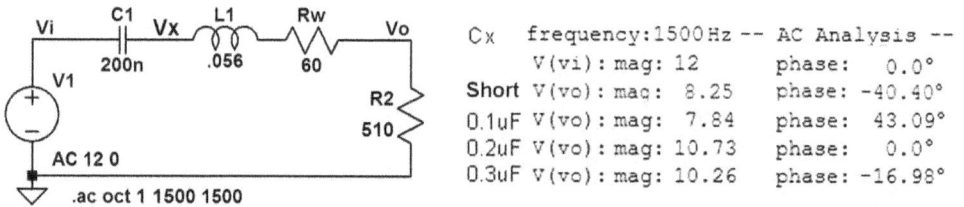

```
Cx    frequency:1500Hz -- AC Analysis --
         V(vi): mag: 12        phase:   0.0°
Short V(vo): mag:  8.25        phase: -40.40°
0.1uF V(vo): mag:  7.84        phase:  43.09°
0.2uF V(vo): mag: 10.73        phase:   0.0°
0.3uF V(vo): mag: 10.26        phase: -16.98°
```

Figure 2-6

The simulation was run four times, once for each value of Cx. Voltage at node Vx can be used to calculate the reactive power in the inductor and capacitor. The reactive power of the capacitor will equal the reactive power of the inductor when the power factor equals one.

Experiment 2c: Parallel Compensation

This experiment demonstrates power factor compensation of a load using a parallel connected compensation component.   The circuit block diagram in Figure 2-7 below shows the voltage source, V_S, with internal resistance, R_S, connected to a load whose apparent power is $S_1 + S_2$.

Power factor compensation of the load is accomplished by connecting a component in parallel with the load so that its reactive VARs, S_3, cancel the reactive VARs of $S_1 + S_2$.

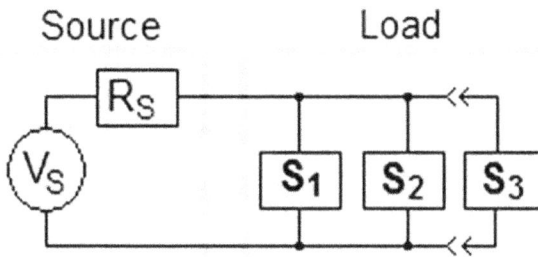

Figure 2-7

Voltages in the series-parallel RL circuit are measured. The power associated with the inductance and resistance is calculated from the voltage measurements. Parallel compensation is added to the circuit and the compensated circuit is compared to the uncompensated one.

Equipment and Parts

Function Generator, Oscilloscope, and Breadboard.
C = 200nF, 5% (two 100nF). R_S = 560, ¼W, 5%. R2:1K, ¼W, 5%.
L: 56mH, 5%. Mouser part #434-02-563J (see appendix 2).

Measure the resistance, R_S, of the inductor.

For greater accuracy, measure the values of R_2, L_1, and C.

R_S _____ R_1 _____ R_2 _____ L_1 _____   C _____

Procedure

1. Connect the circuit in Figure 2-8 <u>without the capacitor, C</u>. This connection will be referred to as the uncompensated circuit.

2. Connect oscilloscope channel 1 to measure **Vi** and oscilloscope channel 2 to measure **Vo**. Set the trigger to channel 1.

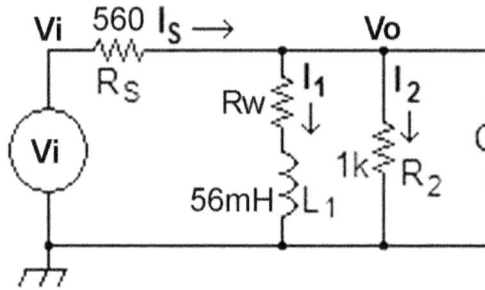

Figure 2-8

3. Set the generator, **Vi**, to produce a 12V peak-to-peak, 1500Hz sine wave with no offset. Set channels 1 and 2 of the oscilloscope to get a display similar to that shown in Figure 2-9.

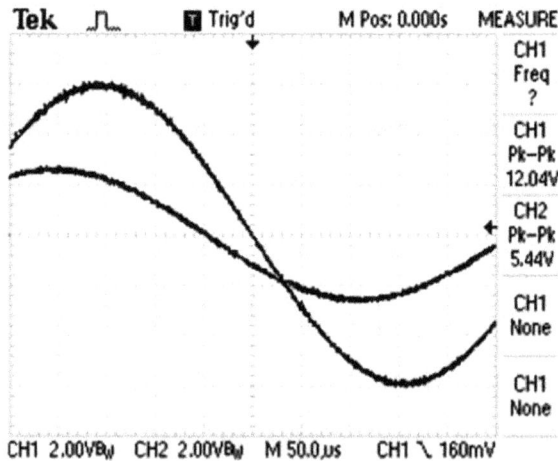

Figure 2-9

Both traces are centered vertically and the reference voltage, **Vi**, crosses zero at the horizontal center of the screen.

4. Adjust the oscilloscope's time base to measure the time difference between the zero crossings of **Vi** and **Vo** most accurately. In figure 2-9, **Vo** crosses zero volts about 44 microseconds before **Vi**. It is leading **Vi** so its angle is positive.

The angle of **Vo** in figure 2-9 $= \dfrac{44}{667} 360 = 23.75^0$.

5. Set up a spreadsheet as shown below..

	A	B	C	D	E	F	G	H	I
1	Circuit	t mS	|Vo|	θo	|I| mA	θi	|Si| mVA	Pi mW	Po mW
3	Un-Comp								
4	Comp								

6. Measure the peak-to-peak amplitude of **Vo,** convert it to RMS, and enter the value into spreadsheet cell C3.

7. Measure the time difference between the zero crossings of **Vi** and **Vo**.

   If available, use "cursors" to measure the time difference. Record the results into spreadsheet cell B3. The time difference is positive if **Vo** crosses zero before **Vi**, and negative if **Vo** crosses zero after **Vi**.

8. Connect the 0.2μF (two .1μF in parallel) compensating capacitor, C, into the circuit. Check that the generator's output is still 12 V peak-to-peak.

9. Measure the magnitude of **Vo,** convert it to RMS, and enter the value into spreadsheet cell C4.

10. Measure the time difference between the zero crossings of **Vi** and **Vo.** Record the results into spreadsheet cell B4 (time is negative if **Vo** crosses zero after **Vi**).

Analysis

1. The spreadsheet can calculate the value for column D using the measurements in column B.

   Cell D2: =(B2/667)*360.

2. Magnitude of **I** in column E and angle in column F is calculated by:

$$I=\frac{4.24\angle 0-|\text{Vo}|\angle\theta o}{\text{Rs}}=|\text{I}|\angle\theta i.$$

3. Enter result into spreadsheet cells E and F. Enter current in mA units.

	A	B	C	D	E	F	G	H	I
1	**Circuit**	**t µS**	**|Vo|**	**θo**	**|I| mA**	**θi**	**|Si| mVA**	**Pi mW**	**Po mW**
3	**Un-Comp**	61	2.19	33.2	4.8	-26.45	20.352	18.22	4.796
4	**Comp**	0.2	2.61	0.1	2.9	0	12.296	12.3	6.812

G3: =4.24*E3   H3: =G3*cos(3.14*F3/180)   I3: = C3*C3/1

|Si| and Pi are in mW because I is in mA. Po is the power delivered to R2, the 1K resistor..

4. Show that the 200nF capacitor is approximately the required capacitor value to compensate the circuit at 1500Hz.

5, Calculate the efficiency of the compensated and uncompensated circuit in delivering power to R2. %Eff = (P/Ps) × 100%.

6. Simulate the circuit using AC sweep analysis. Use your measured component values.

7. Explain why the compensated circuit delivers more power to the load (R2) than the uncompensated one.

8. Calculate the theoretical power factor of the uncompensated and of the compensated circuit at 1500 Hz.

9. Calculate the power factor of the uncompensated circuit and of the compensated circuit at 1500 Hz, using measured voltages at node N2. Compare the results.

LTspice Example

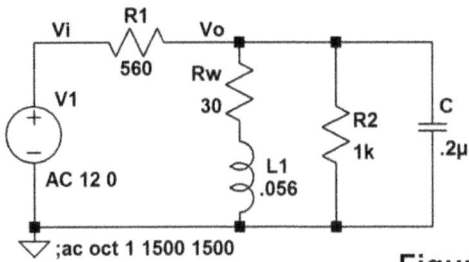

frequency: 1500Hz -- AC Analysis --
Uncompensated

V(vi): mag: 12	phase: 0°
V(vo): mag: 6.20	phase: 33.13°

Compensated

V(vi): mag: 12	phase: 0°
V(vo): mag: 7.40	phase: 0.07°

Figure 2-10

LTspice Example

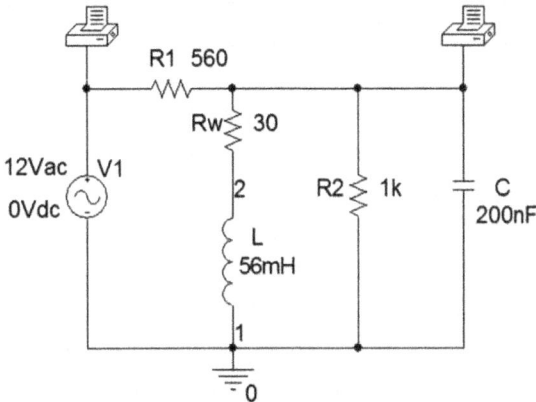

Compensated Results

Vo = 7.407E+00 Volts at 7.217E-02 Degrees.

The printers, "VPRINT1", are in the "SPECIAL" library.   Double click on each printer to open its property editor. Set the following in the "Property Editor" : *AC* = ok, *MAG* = ok, and *PHASE* = ok

Select *Analysis Type*: AC Sweep/Noise.   *AC Sweep Type*: Linear.

Start frequency: 1500Hz. *Stop frequency*: 1500Hz. *Total Points*: 1.

Chapter 3: Single Phase Transformers

Transformers are used to transform voltages, match impedances, and provide isolation. Transformers in this book will be applied to voltage transformation. They will be modeled as ideal transformers with series "winding" resistances, as shown in Figure 3-1.

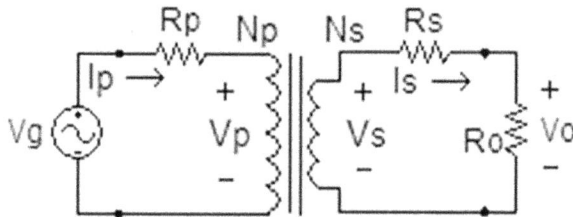

Figure 3-1

The voltage, current, and impedance transformation equations for an ideal transformer are given below.

$$\frac{V_P}{V_S} = \frac{N_P}{N_S} \qquad \frac{I_P}{I_S} = \frac{N_S}{N_P} \qquad \left(\frac{Z_P}{Z_S}\right) = \left(\frac{N_P}{N_S}\right)^2$$

Transformer inductance, capacitance, and hysteresis effects will be assumed to be negligible.

The transformer's primary impedance will be a reflection of its secondary impedance.

Calculation Example

Consider the transformer circuit in figure 3-1. Let's say it's a step-down transformer with the following specifications:

Turns ratio = 5 to 1. Ro = 100Ω, Rs = 10Ω, and Rp = 40Ω.

An input voltage, Vg, of 600VAC (RMS) is applied.

Calculate the primary current, Ip, secondary currents, Is, output voltage, Vo, and the transformers efficiency.

First calculate the primary impedance:

$$\left(\frac{Z_P}{Ro+Rs}\right)=\left(\frac{Z_P}{100+10}\right)=\left(\frac{5}{1}\right)^2$$

$$Zp=25(100+10)=2750\Omega$$

Next calculate the primary current and voltage:

$$Ip=\frac{Vg}{Zp+Rp}=\frac{600}{2750+40}=0.215A$$

$$Vp=Vg-Ip\,Rp=600-0.215(40)=591.4V$$

Next calculate the secondary voltage and current:

$$\frac{Vp}{V_S}=\frac{591.4}{V_S}=\frac{5}{1}\quad\Rightarrow\quad Vs=\frac{591.4}{5}=118.3V$$

$$Is=\frac{Vs}{Ro+Rs}=\frac{118.3}{110}=1.075A$$

$$\text{Alternately, }\frac{Ip}{I_S}=\frac{.215}{I_S}=\frac{1}{5}\quad\Rightarrow\quad Is=5(.215)=1.075A$$

Calculate the input and output powers and power transfer efficiency:

$$P_{IN}=Vg\,Ip=600(.215)=129W\qquad P_{OUT}=Is^2Ro=(1.075)^2\,100=115.6W$$

$$\text{Efficiency}=\frac{115.6W}{129W}100\%=89.6\%$$

Experiment 3a: Step-Down Transformer

The operating characteristics of a small step-down transformer will be measured and analyzed.

Equipment and Parts

Function Generator, Oscilloscope, DMM, and Breadboard.
Transformer, 500Ω CT to 500Ω CT, 400mW. Refer to appendix 2.
  Recommended: ZICON 42TU500-RC (from Mouser Electronics)
Resistors: Rg = 10Ω, ¼ watt, 5%. Ro = 470Ω, ¼ watt, 5%.

Measure the resistance, Rp of the primary winding and the resistance, Rs, of one-half of the secondary winding.

Rp = _____   Rs = _____

For greater accuracy, measure the values of Rg, and Ro.

Rg = _____   Ro= _____

Procedure

1. Connect the circuit in Figure 3-2. Connect oscilloscope channel 1 to measure Vg and channel 2 to measure Vo. Set the generator to produce a 400Hz, 12V p-p sine wave with no offset. Set the oscilloscope to trigger on channel 1. The transformer pin numbers are for the recommended transformer.

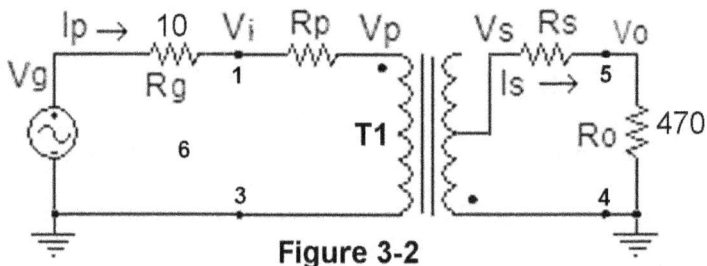

Figure 3-2

2. Measure and record the magnitude and phase of the voltage Vo.

   Vo(mag.) _____V p-p   Vo(angle) _____ Degrees

3. Measure and record the magnitude and phase of the voltage Vi with channel 2 of the oscilloscope..

Vi(mag.) _____V p-p.   Vi(angle) _____ degrees.

Analysis

1. Use the measurements in procedure steps 2 and 3 to calculate the primary current, secondary current, power supplied by the source, power dissipated by the 470Ω load, and the efficiency of the power transfer from the source to the load.

2. Explain the results for the phase angles between the source, Vg, and the voltages Vi and Vo in steps 2 and 3.

3. Simulate the circuit of Figure 3-2 and compare your results to the calculated results from analysis step 1.

LTspice Simulation Example: Step-down transformer

In LTspice, a transformer is modeled with coupled inductors. A mutual inductance statement is placed as a SPICE directive on the schematic. To enter a SPICE directive, click on "Edit" in the menu bar and select "SPICE Directive". The dialog box shown in Figure 3-3 will open.

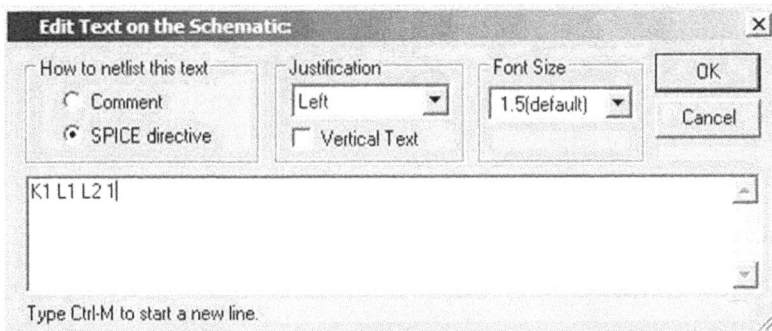

Figure 3-3

The diagram in Figure 3-3 includes the directive: K1 L1 L2 1. K1 is the coupling coefficient between L1 and L2 and is set to 1. Inductors must have a phasing dot. LTspice adds the phasing dots automatically when coupling directive is added.

The transformer turns ratio is equal to the square root of the inductance ratio. First the primary inductance needs to be approximated Then the secondary inductance is calculated from the turns ratio.

The transformer configuration in Figure 3-4 has an inductance ratio of 4 to 1 and a turns ratio of 2 to 1.

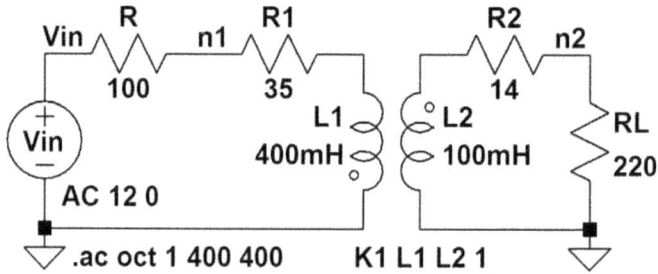

Figure 3-4

Simulation was set to "AC Analysis" at 400Hz. The analysis results below show a 180 degree phase difference between the primary and secondary. This was expected from the inductor phasing dots.

```
       --- AC Analysis ---
frequency:      400            Hz
V(n001):    mag:   10.4159 phase:     6.69362°
V(n1):      mag:   10.8115 phase:     4.77148°
V(vin):     mag:        12 phase:          0°
V(n002):    mag:   5.20795 phase:   -173.306°
V(n2):      mag:   4.89637 phase:   -173.306°
```

Suggested exercise

Calculate the values of V(n1) and V(n2) and compare the results to the simulation above.

Calculate the power transfer efficiency of the transformer using the ratio of the power delivered by the secondary (at node n2) to the power supplied to the primary (at node n1).

Single-Phase Power Systems

In the United States 60Hz AC Electric power is typically supplied to a residence by a 240VAC center tapped transformer. This transformer may be supplied by one phase of a three-phase distribution system. It may be on a pole or underground. The transformer's center tap is connected to earth ground. Figure 3-5 is a simplified diagram showing a step-down transformer, power meter, and breaker panel.

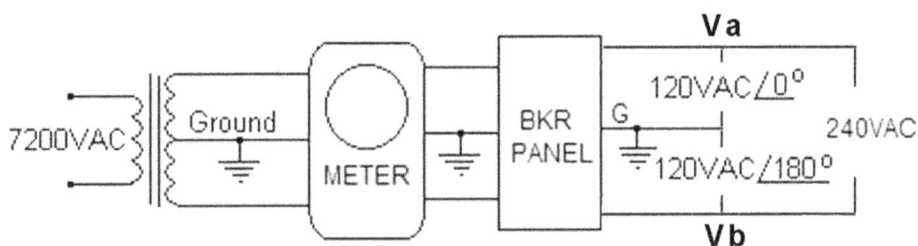

Figure 3-5

This transformer provides 120VAC on each side of the tap (**Va** to **G** and **Vb** to **G**) and 240VAC across the entire secondary (**Va** to **Vb**). Note that **Vb** to **G** is 180° out of phase with **Va** to **G**.

The actual voltage available at a 120VAC or 240VAC outlet is current dependent. The Voltage dropped across the transformer windings and distribution wires is proportional to current flow. This results in a power loss in the wiring resistance that is proportional to the square of the current. Other types of losses occur in power transformers which typically range from 2% to 4% of the power being supplied.

This chapter's Experiment 4 demonstrates some of the characteristics of a center tapped transformer distribution system. However, the voltages are scaled down so that the experiment may be safely performed on a breadboard. It uses the same transformer as used in Experiment 3.

The transformer is powered by a 400Hz sine wave from a signal generator. 400Hz is used here because the small 400mW transformer is less efficient at 60Hz. 400Hz is a common power distribution frequency on aircraft and marine vehicles because the transformers can be lighter. Alternately, a 12VAC center tapped line operated transformer could be used to do this experiment.

Experiment 3b: Center Tapped Transformer

The operating characteristics of a center tapped transformer power source will be measured. The transformer's input power, output power. and efficiency will be determined.

Equipment and Parts

Function Generator, Oscilloscope, DMM, and Breadboard.
Transformer, 500Ω CT to 500Ω CT, 400mW. Refer to appendix 2.
Recommended: ZICON 42TU500-RC (from Mouser Electronics)
Resistors: Two 10Ω, three 470Ω, all ¼ watt, 5%.

Procedure: Part 1, Balanced Load

1. Connect the circuit in Figure 3-6. Set the generator to produce a 400Hz, 12V p-p sine wave with no offset. Connect oscilloscope channel 1 to measure Vp and channel 2 to measure Vs. Trigger on channel 1.

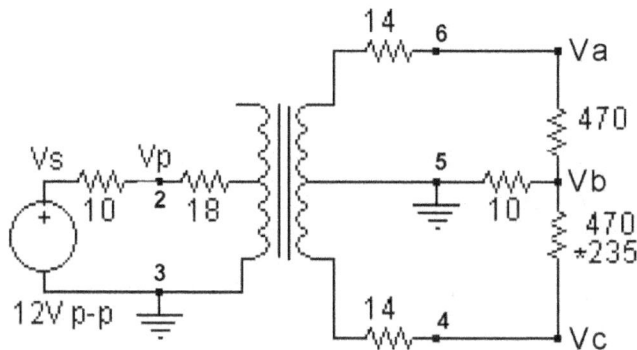

Figure 3-6

2. Measure and record the magnitude Vs and Vg.

    Vs _____ V p-p   Vg _____ V p-p

3. Measure and record the magnitude and phase angle of the voltage Va with channel 2 of the oscilloscope.

    Va(mag.) _____ V p-p.   Va(angle) _____ degrees.

44

4. Measure and record the magnitude and phase angle of the voltage Vb with channel 2 of the oscilloscope.

Vb(mag.)_____V p-p.   Vb(angle)_____ degrees.

5. Measure and record the magnitude and phase angle of the voltage Vc with channel 2 of the oscilloscope.

Vc(mag.)_____V p-p.   Vc(angle)_____ degrees.

Procedure: Part 2, Un-Balanced Load

1. Connect a 470 ohm resistor in parallel with the resistor connected between Vb and Vc.

2. Measure and record the magnitude Vs and Vg.

Vs_____V p-p   Vg_____ V p-p

3. Measure and record the magnitude and phase angle of the voltage Va with channel 2 of the oscilloscope.

Va(mag.)_____V p-p.   Va(angle)_____ degrees.

4. Measure and record the magnitude and phase angle of the voltage Vb with channel 2 of the oscilloscope.

Vb(mag.)_____V p-p.   Vb(angle)_____ degrees.

5. Measure and record the magnitude and phase angle of the voltage Vc with channel 2 of the oscilloscope.

Vc(mag.)_____V p-p.   Vc(angle)_____ degrees.

Note: All of the voltage measurements in this experiment are in peak-to-peak units. Power must be calculated using RMS units. A spreadsheet may be used to convert the peak-to-peak units to RMS units and to calculate the transformer's power input and output.

Analysis

1. Enter results into a spreadsheet. Calculate input power, output power and efficiency for parts 1 and 2. Refer to the example below:

	A	B	C	D	E	F	G	H	I	J	K
1	Part	Vs p-p	Vp p-p	Ip p-p	Va p-p	Vb p-p	Vc p-p	In p-p	Pin	Pout	%Eff.
2	1	12	11.57	0.043	10.43	0	10.43	0	0.062	0.058	93.05
4	2	12	11.4	0.06	9.95	0.188	8.69	0.019	0.086	0.061	71.85

Equations in cells: D2: =(B2-C2)/10   H2: =F2/10   I2: =(C2*D2)/8

J2: =(E2+G2)^2/(8*940)   K2:=(J2/I2)*100

2. Explain the results for the phase angles between the source, Vp, and the voltages Va and Vc.

3. Simulate the circuit of Figure 3-6 and compare your results to the to your measurements.

4. Calculate the output power for parts 1 and 2 using the ideal transformer model and compare the calculated results to your measurements and simulation.

LTspice Example

The circuit in figure 3-7 below represents the circuit of part 2 of this experiment. It was simulated using the ideal transformer model and including the transformers primary and secondary resistances.

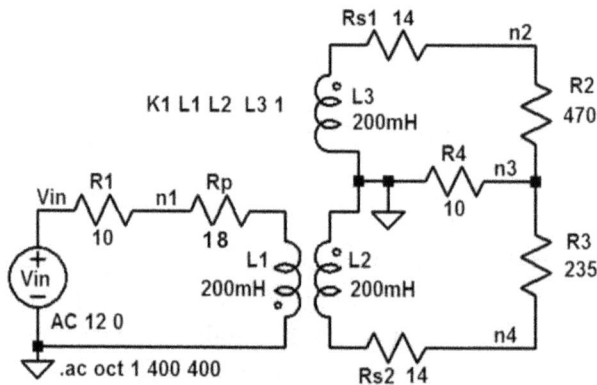

Figure 3-7

Analysis Results

```
--- AC Analysis --- frequency: 400Hz
V(vin): mag: 12.0      phase:    0.0°
V(n1):  mag: 11.37     phase:    0.8°
V(n2):  mag:  9.94     phase: -177.2°
V(n3):  mag:  0.18     phase:    2.7°
V(n4):  mag:  9.68     phase:    2.7°
```

In the circuit in figure 3-8 below the 10 ohm resistor between node n3 and ground is removed. It was simulated using the ideal transformer model and including the transformers primary and secondary resistances.

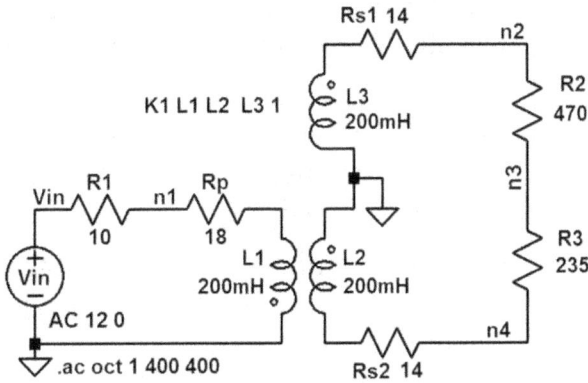

Figure 3-8

Analysis Results

Note below that the voltage at node n3 is 3.33 volts. This causes the voltage across R2 to increase to 13.33 volts and across R3 to decrease to 6.67 volts.

```
frequency: 400Hz --- AC Analysis ---
V(n1):  mag: 11.42     phase:    0.89°
V(vin): mag: 12.00     phase:    0.00°
V(n2):  mag: 10.00     phase: -177.23°
V(n3):  mag:  3.33     phase:    2.76°
V(n4):  mag: 10.00     phase:    2.76°
```

Experiment 3c: Audio Output Transformer

An AC voltage source with a source impedance of 1000Ω is first connected directly to an 8.2Ω resistor, and the power delivered to it is measured. Next, the same load resistance is connected to the source through an impedance matching transformer, and the power delivered to the 8.2Ω resistor is again measured.

The primary and secondary voltages and currents will be determined and compared to the voltage transformation properties of the transformer.

An audio output transformer is used which is designed to match a high impedance audio amplifier output to a low impedance loudspeaker. An 8Ω loudspeaker will replace the 8.2Ω resistor, and measurements will be repeated. This will also provide an audible demonstration of the effectiveness of the transformer.

In the second part of this exercise the transformer primary will be connected as an "auto-transformer", and its voltage and impedance matching properties will be explored.

This lab experiment demonstrates that the voltage, current, and impedance transformation properties of an iron core transformer approximate that of an ideal transformer.

Equipment and Parts

Function Generator, Oscilloscope, DMM, and Breadboard.
Audio Transformer, 1000Ω, center tapped, to 8Ω, 200mW.
Loudspeaker, 8Ω, 200mW minimum, (2 to 4 inch).
Resistors: R1: 8.2, R2: 1000, R3: 220, R4: 47, R5: 470, all ¼ watt, 5%.

Procedure Part 1: Impedance Transformation

1. Measure the values of the resistors for use in your analysis.

    R1 _____   R2 _____   R3 _____

    R4 _____   R5 _____

2. Connect the circuit on the right. Connect channel 1 of the oscilloscope to Vg, the function generator. Connect channel 2 of the oscilloscope to measure Vo. Set the oscilloscope to trigger on channel 1.

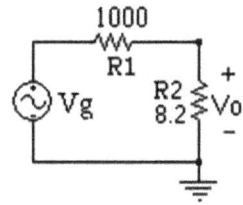

Set the function generator to produce a 10V p-p, 1000Hz sine wave with no offset.

3. Measure and record the magnitude and phase of the voltage Vo.

Vo(mag) _____ V p-p   Vo(angle) _____ Degrees

4. Connect the circuit on the right. Connect channel 1 of the oscilloscope to Vg, the function generator. Connect channel 2 of the oscilloscope to measure Vo.

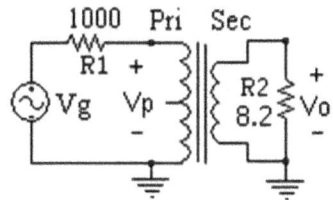

Set the oscilloscope trigger to channel 1.

. Set the function generator to produce a 10V peak-to-peak, 1000Hz sine wave with no offset.

5. Measure and record the magnitude and phase of the voltage Vo.

Vo(mag) _____ V p-p.   Vo(angle) _____ degrees.

6. Measure and record the magnitude and phase of the voltage Vp.

Vp(mag) _____ V p-p.   Vp(angle) _____ degrees.

7. Measure and record the resistance of the loudspeaker.

Sp _____ ohms.

8. Connect the speaker circuit on the right. Connect channel 1 of the oscilloscope to Vg, the function generator. Connect channel 2 of the oscilloscope to measure Vo.

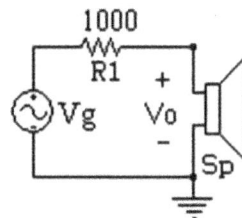

Set the oscilloscope to trigger on channel 1.

49

Set the function generator to produce a 10V p-p, 1000Hz sine wave with no offset.

9. Measure and record the magnitude and phase of Vo.

Vo(mag) _____V p-p. Vo(angle) _____degrees.

10. Connect the speaker circuit on the right. Connect channel 1 of the oscilloscope to Vg, the function generator.

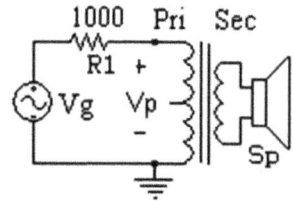

Connect channel 2 of the oscilloscope to measure Vo.

Set the oscilloscope to trigger on channel 1. Set the function generator to produce a 10V p-p, 1000Hz sine wave with no offset.

11. Measure and record the magnitude and phase of Vo.

Vo(mag) _____V p-p.   Vo(angle) _____ degrees.

12. Measure and record the magnitude and phase of the voltage Vp.

Vp(mag) _____V p-p.   Vp(angle)_____ degrees.

Part 2: Autotransformer

1. Connect the circuit on the right. R_L = 47Ω. Connect   channel 1 of the oscilloscope to Vg, the function generator.

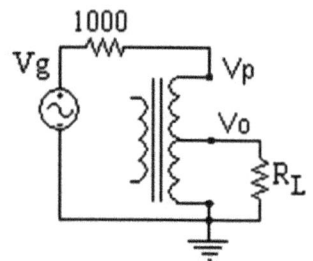

Connect channel 2 of the oscilloscope to measure Vo.

Set the oscilloscope to trigger on channel 1. Set the function   generator to produce a 10V p-p, 1000 Hz sine wave with no offset.

2. Measure and record the magnitude and phase of the voltage Vo.

Vo(mag) _____V p-p.   Vo(angle)_____ degrees.

3. Measure and record the magnitude and phase of the voltage Vp.

    Vp(mag)_____V p-p.   Vp(angle)_____degrees.

4. Repeat step 1 with R_L = 220Ω.

5. Measure and record the magnitude and phase of the voltage Vo.

    Vo(mag)_____V p-p.   Vo(angle)_____degrees.

6. Measure and record the magnitude and phase of the voltage Vp.

    Vp(mag)_____V p-p.   Vp(angle)_____degrees.

7. Repeat step 1 with R_L = 470Ω.

8. Measure and record the magnitude and phase of the voltage Vo.

    Vo(mag)_____V p-p.   Vo(angle)_____degrees.

9. Measure and record the magnitude and phase of the voltage Vp.

    Vp(mag)_____V p-p.   Vp(angle)_____degrees.

Analysis, Part 1

1. Use the measurements in step 3 to calculate power delivered to the 8.2Ω load resistor.   Compare measured power to the theoretically expected power. Calculate the ratio of the power delivered to the load resistor to the power delivered by the source.

2. Use the measurements in step 5 to calculate power delivered to the 8.2Ω load resistor. Compare measured to theoretically expected power. Calculate the ratio of load resistor power to   power delivered by the source (also use results of step 6).

3. Use the measurements in steps 5 and 6 to calculate the voltage and current transformation ratios of the transformer, and compare the measured values to the theoretically expected values.

4. Explain the results for the phase angles between the source, Vg, and the voltages Vp and Vo in steps 5 and 6.

5. Optional and challenging:

   Refer to the circuit on the right. Measure the inductance, Lp and Ls, and resistance, Rp and Rs, of the primary and secondary windings.

   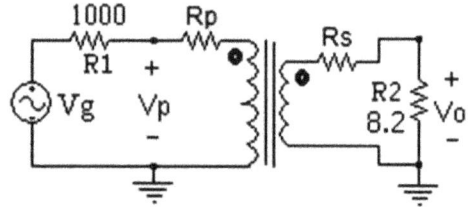

   Calculate the voltage, Vo, across the 8.2 ohm load resistor by including the transformer's winding resistances, Rp and Rs.

   Use the mesh current method and assume a coupling coefficient of 0.95. Compare your calculated results to your measured results

Analysis, Part 2

1. Use measurements in part 2, steps 2 and 3, to calculate the power delivered to 47Ω load resistor. Compare the measured power to the theoretically expected power.

   Calculate the ratio of the power delivered to the load resistor to the power delivered by the source.

2. Use the measurements in part 2, steps 5 and 6, to calculate the power delivered to 220Ω load resistor. Compare the measured power to the theoretically expected power.

   Calculate the ratio of the power delivered to the load resistor to the power delivered by the source.

3. Use the measurements in part 2, steps 8 and 9, to calculate the power delivered to 470Ω load resistor. Compare measured power to the theoretically expected power.

   Calculate the ratio of the power delivered to the load resistor to the power delivered by the source.

4. Compare and comment on the results of the ratio of power delivered to the load resistor to the power delivered by the source for the 47Ω, 220Ω, and 470Ω load resistors.

Transformer Simulation Notes

You need to approximate the inductance of the primary winding (300mH was used in this simulation).

The secondary inductance is calculated from the turns ratio which is equal to the square root of the inductance ratio.

Part 1: LTspice

In LTspice, a transformer is modeled with coupled inductors with a mutual inductance statement placed as a SPICE directive on the schematic.

The diagram below includes the directive: K1 L1 L2 1. K1 is the coupling coefficient between L1 and L2 and is set to 1. Inductors must have a phasing dot.

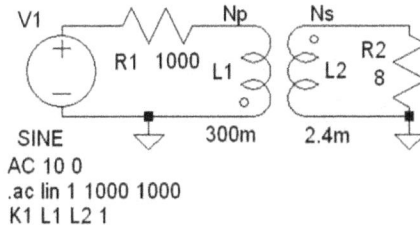

```
        --- AC Analysis ---

frequency: 1000 Hz
V(np):  mag:     4.83287 phase:  14.8561°
V(ns):  mag:    0.432265 phase:-165.144°
V(n001) mag:          10 phase:       0°
I(R2):  mag:   0.0540331 phase:-165.144°
I(R1):  mag:  0.00547085 phase:  166.909°
I(V1):  mag:  0.00547085 phase:  166.909°
```

Part 2: PSpice / Auto-transformer.

The evaluation version of PSpice has the linear transformer, XFRM_LINEAR, that can be used to simulate this lab. Double click on the transformer to open the property editor. There are three parameters that can be set for the simulation: COUPLING, L1_VALUE, and L2_VALUE.

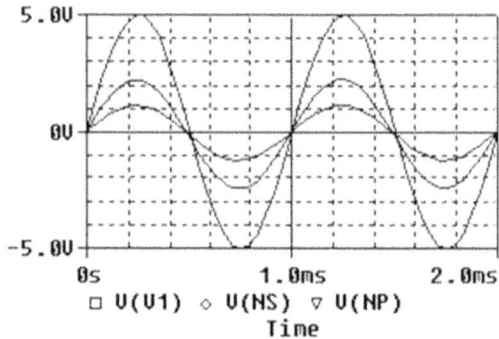

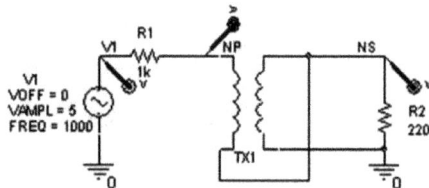

TX1 PROPERTY EDITOR:
COUPLING=1, L1_VALUE=300m, L2_VALUE=300M

Chapter 4: Three-Phase Power Systems

Three-phase power is used to transmit and distribute electricity by electrical grids world wide. Three-phase systems are usually more efficient and economical than single-phase systems. Three-phase motors and generators operate with less vibration because the average power transferred to a balanced three-phase load is constant and not a function of time. The design and construction of electric machines is simplified because it is easy to generate a rotating magnetic field.

Electric Power distributed to residential customers is single-phase, typically derived from one phase of a three-phase source. In the USA and some other countries customers receive single-phase 240VAC and 120VAC at a frequency of 60Hz. In Europe and some other countries customers receive single-phase 230VAC at a frequency of 50Hz.

Electric Power distributed to industrial customers in the USA is typically three-phase, 208VAC line to line, and 120VAC line to neutral, at a frequency of 60Hz. Higher three-phase voltages are also available.

A three-phase voltage source generates three sine waves whose phase angles are 120 degrees apart. Refer to figure 4-1 below. Three phasors are shown, **Va** at 0 degrees, **Vb** at 120 degrees, and **Vc** at 240 degrees. The graph shows the magnitude of the vertical projection of each phasor as a function of time as it rotates counter clockwise.

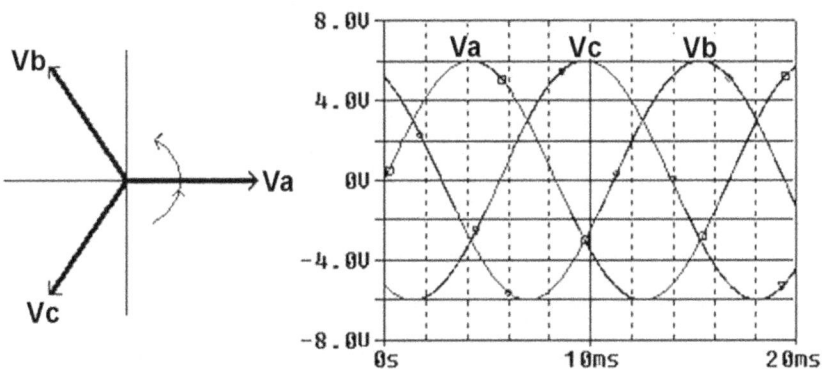

Figure 4-1

Above: **Va** = 12∠0⁰, **Vb** = 12∠120⁰, **Vc** = 12∠240⁰.

Va, **Vb**, and **Vc** are referred to as "line to neutral" voltages. The voltages between the lines are referred to as "line to line" voltages. The magnitudes of the line to line voltages, **Vab**, **Vbc**, and **Vca**, are $\sqrt{3}$ larger than the magnitudes of the phase-to-neutral voltages. For example:

$$\mathbf{Vab} = \mathbf{Va} - \mathbf{Vb} = 12\angle 0^0 - 12\angle 120^0 = 20.28\angle - 30^0$$

$$\mathbf{Vbc} = \mathbf{Vb} - \mathbf{Vc} = 12\angle 120^0 - 12\angle 240^0 = 20.28\angle 90^0$$

$$\mathbf{Vca} = \mathbf{Vc} - \mathbf{Va} = 12\angle 240^0 - 12\angle 0^0 = 20.28\angle - 150^0$$

A three-phase voltage source may be connected in a "wye" or "delta" configuration. Similarly, three-phase loads may be connected in a wye or "delta" configuration. The "wye" is also called "star".

The load impedances, **Za**, **Zb**, and **Zc**, may be real or complex. Resistors will be used for most of the impedances in this chapter to simplify the schematic diagrams and calculations. Complex impedances will be given in rectangular or polar form, such as: $\mathbf{Z} = R + jX$, or as $\mathbf{Z} = |\mathbf{Z}|\angle\theta$.

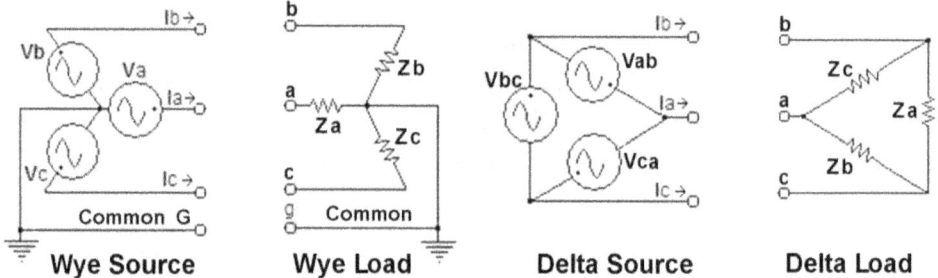

Wye Source   Wye Load   Delta Source   Delta Load

Diagrams of the 3-phase sources and loads are shown above. Note that in the delta source the phases are connected in series in a closed loop. There is no current around the loop because the phasor sum of the voltages is 0.

A three-phase load is called "balanced" if its phase impedances are equal: **Za** = **Zb** = **Zc**. If the phase impedances are not equal the load is called "unbalanced". Most three-phase loads such as motors and heaters are balanced. Power companies distribute three-phase power to approximately balanced loads.

Definitions

Phase voltage is the voltage across each phase of a wye or delta connected source or load. For a wye connection it is measured from a line to neutral. For a delta connection it is measured from line to line.

Line voltage is the voltage measured from line to line. The line voltage equals the phase voltage for each phase of a delta connection.

Phase current is the current in each phase of a wye or delta connected source or load.

Line current is the current in each line connected to the source or load. The line current equals the phase current for each phase of a wye connection.

Example

Wye Load

Delta Load

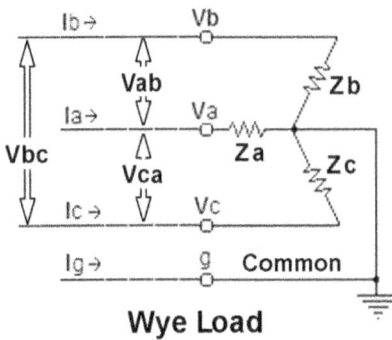

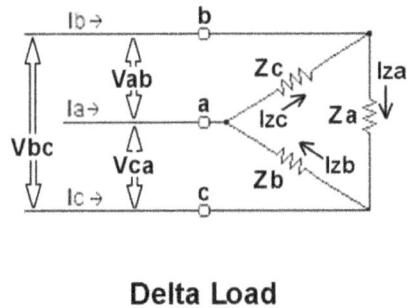

Refer to the wye load diagram above:

Vab, Vbc, and Vca are line to line voltages. The phase voltages, Va, Vb, and Vc are measured with respect to the common ground. Ia, Ib, and Ic are line currents and also phase currents. Ig is the common (also called neutral) current. It is zero in a balanced circuit.

Refer to the delta load diagram above:

Vab, Vbc, and Vca are line to line voltages and also the phase voltages. Ia, Ib, and Ic are line currents. Iza, Izb, and Izc are phase currents.

Three-Phase Calculations

Steady-state three-phase circuit calculations may be done with phasor algebra. Typical techniques include the node voltage method, mesh current method, and Ohm's law.

Phasor calculations, including the simultaneous equations, may be solved on a scientific calculator such as the *TI-89*. Math software such as *Maple, MathCAD*, and *MATLAB* may also be used. Simulation software such as *PSpice* or *LTspice* may be used to check calculations.

The wye-wye and the delta-wye circuits are not readily solved with the mesh current method. This is because the determinant of the coefficient matrix of the mesh equations is equal to 0. Solvers such as Maple, MathCAD, and the TI-89 calculator will not yield a solution. However, the node voltage method may usually be applied successfully to these circuits.

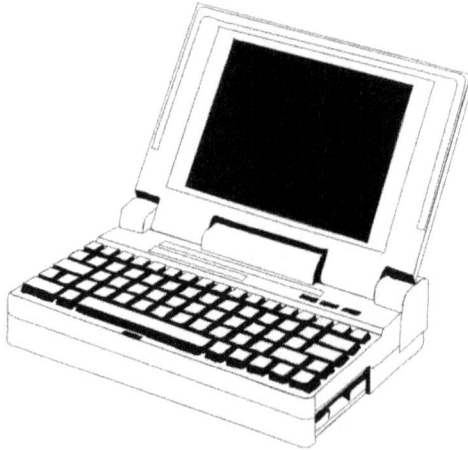

Wye Source to Wye Load

Figure 4-2 below shows a four wire wye source connected to a four wire wye load. This is a simplified diagram. **Zx** represents line impedance. Source impedances are assumed to be negligible. The neutral (ground) wire is not needed if the load is balanced (**Za** = **Zb** = **Zc**).

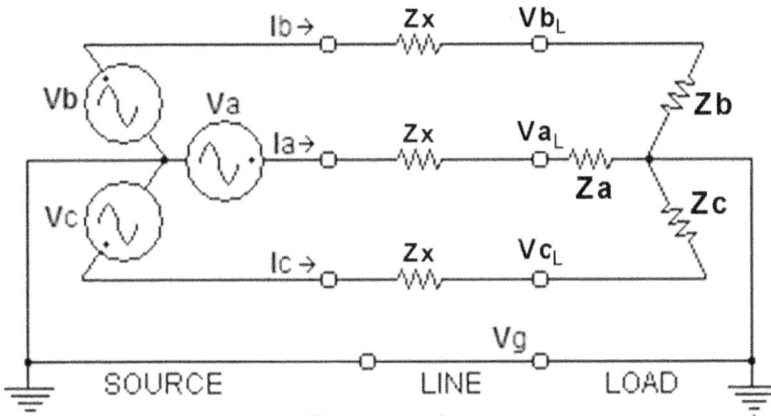

Figure 4-2

Equations for balanced or unbalanced four wire wye-wye circuit or balanced three wire wye-wye circuit:

Line currents can be calculated independently for each phase.

$$Ia = \frac{Va}{Za + Zx} \qquad Ib = \frac{Vb}{Zb + Zx} \qquad Ic = \frac{Vc}{Zc + Zx}$$

The total Power supplied by the source can be calculated by summing the powers for each phase.

$$S = Sa + Sb + Sc = Va(Ia^*) + Vb(Ib^*) + Vc(Ic^*)$$

In polar form, **S** = |**S**|∠θ. The circuit's power factor is: pf = cosθ.

Total Average power is: P = |**S**|cosθ.

Voltage across each phase of the load can be calculated with Ohm's law.

$$\mathbf{Va_L = Ia(Za) \qquad Vb_L = Ib(Zb) \qquad Vc_L = Ic(Zc)}$$

The total power supplied to the load is the sum of the powers supplied to each phase.

$$\mathbf{S_L = Sa_L + Sb_L + Sc_L = Va_L(Ia^*) + Vb_L(Ib^*) + Vc_L(Ic^*)}$$

In polar form, $\mathbf{S_L} = |S_L| \angle \theta_L$. The load power factor is: $pf_L = \cos\theta_L$.

Total Average power supplied to the load is: $P_L = |S_L|\cos\theta$.

Power Transfer Efficiency

The power transfer efficiency, expressed in percent, is equal to the ratio of the average power supplied to the load to the average power supplied by the source times 100.

$$P_T = \frac{P_L}{P} 100\% = \frac{|S_L|\cos\theta_L}{|S|\cos\theta} 100\%$$

Equations for an unbalanced Three wire wye-wye circuit:

The example in figure 4-3 below shows the node voltage method applied to a three wire wye-wye circuit. In a balanced circuit, the voltage **V** at the junction of the load impedances would be zero and the solution would be the same as for the four wire circuit.

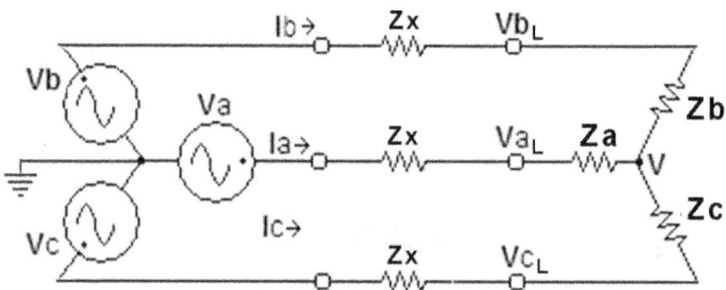

Figure 4-3

Zx represents the line impedance. Source impedances are assumed to be negligible. The node voltage method is used to find the voltage **V** at the junction of the load phase impedances, **Za**, **Zb**, and **Zc**.

All node voltages are measured with respect to the source ground, as shown in figure 5-3. The voltage **V** is calculated using that the sum of the currents, **Ia**, **Ib**, and **Ic**, flowing into node **V** must be 0.

Given: **Va** = 12∠0⁰ Vrms,   **Vb** = 12∠120⁰Vrms,   **Vc** = 12∠-120⁰Vrms,

    Za = **Zb** = 100, **Zc** = 50. **Zx** =10.

$$\frac{Va\text{-}V}{Za+Zx}+\frac{Vb\text{-}V}{Zb+Zx}+\frac{Vc\text{-}V}{Zc+Zx}=0 \quad \Rightarrow \quad \frac{12\text{-}V}{110}+\frac{12\angle120\text{-}V}{110}+\frac{12\angle-120\text{-}V}{60}=0$$

The result shows that **V** = 2.61∠-120⁰ volts. If the circuit were balanced the result would have been 0.0 volts. In this circuit Ia, Ib, and Ic are the line currents and also the phase currents. They are calculated below.

$$\mathbf{Ia}=\frac{\mathbf{Va\text{-}V}}{\mathbf{Za+Zx}}=.123\angle9.6^0 \quad \mathbf{Ib}=\frac{\mathbf{Vb\text{-}V}}{\mathbf{Zb+Zx}}=.123\angle110.4^0 \quad \mathbf{Ic}=\frac{\mathbf{Vc\text{-}V}}{\mathbf{Zc+Zx}}=.157\angle-120^0$$

Total power supplied by the source is:

$$S=Va(Ia^*)+Vb(Ib^*)+Vc(Ic^*)$$
$$S=(12\angle0^0)(.123\angle-9.6^0)+(12\angle120^0)(.123\angle-110.4^0)+(12\angle-120^0)(.157\angle120^0)$$
$$S=1.476\angle-9.6^0 \;+\; 1.476\angle9.6^0 \;+\; 1.884\angle0^0 \;=\; 4.795 \text{ Watts}$$

Total power delivered to the load is:

$$S_L=Ia(Ia^*)Za+Ib(Ib^*)Zb+Ic(Ic^*)Zc,$$
$$S_L=(.123^2)100+(.123^2)100+(.157^2)50=4.258 \text{Watts..}$$

Power transfer efficiency is:

$$P_T=\frac{P_L}{P}100\%=\frac{|S_L|cos\theta_L}{|S|cos\theta}100\%=\frac{4.258}{4.795}100\%=89\%$$

Eleven percent of the source power was lost in the transmission line resistance

Balanced three wire wye-wye circuit:

The magnitudes of the phase voltages and currents are equal.

$$|Va| = |Vb| = |Vc| = |V| \quad \Rightarrow \quad |Ia| = |Ib| = |Ic| = |I|$$

$$Ia = \frac{|V| \angle 0^0}{|Z + Zx| \angle \theta^0} = |I| \angle (0^0 - \theta^0)$$

$$Ib = \frac{|V| \angle 120^0}{|Z + Zx| \angle \theta^0} = |I| \angle (120^0 - \theta^0)$$

$$Ic = \frac{|V| \angle -120^0}{|Z + Zx| \angle \theta^0} = |I| \angle (-120^0 - \theta^0)$$

Total Power delivered to the load:

$$S_L = Va_L (Ia^*) + Vb_L (Ib^*) + Vc_L (Ic^*) = |V_L||I| \angle \theta^0 + |V_L||I| \angle \theta^0 + |V_L||I| \angle \theta^0$$

$$S_L = 3|V_L||I| \angle \theta^0 \qquad P_L = 3|V_L||I| \cos \theta^0$$

Note that the magnitudes of the load's phase voltages are equal:
$$|Va_L| = |Vb_L| = |Vc_L| = |V_L| = |I| Z$$

Using line to line voltage, where $|V_{LL}| = \sqrt{3}|V|$:

$$S = 3 \frac{|V_{LL}|}{\sqrt{3}} |I| \angle \theta^0 = \sqrt{3} |V_{LL}||I| \angle \theta^0.$$

LTspice Example

LTspice simulation of circuit yielded the results in figure 4-4 below.

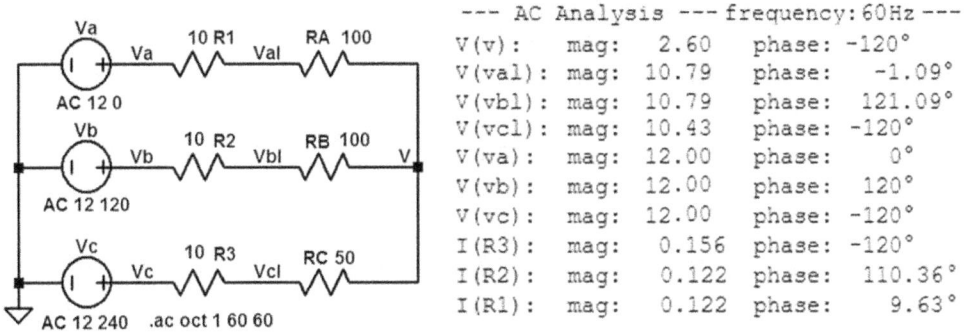

```
--- AC Analysis --- frequency: 60Hz ---
V(v):    mag:   2.60    phase: -120°
V(val):  mag:  10.79    phase:  -1.09°
V(vbl):  mag:  10.79    phase: 121.09°
V(vcl):  mag:  10.43    phase: -120°
V(va):   mag:  12.00    phase:    0°
V(vb):   mag:  12.00    phase:  120°
V(vc):   mag:  12.00    phase: -120°
I(R3):   mag:   0.156   phase: -120°
I(R2):   mag:   0.122   phase: 110.36°
I(R1):   mag:   0.122   phase:   9.63°
```

Figure 4-4

These results agree with the previous unbalanced three wire circuit calculations.

Power supplied by the source is calculated below. The circuit contains no reactances, so the phase angle of **S** is 0 and the power factor is 0.

$$S=(12\angle0^0)(.122\angle-9.63^0)+(12\angle120^0)(.122\angle-110.4^0)+(12\angle240)(.156\angle120^0).$$
$$S=1.46\angle-9.6^0+1.46\angle9.6^0+1.87\angle0^0=4.75\angle0^0 \text{ Watts}$$

The power supplied to the load, S_L, is calculated below. The phase angle of S_L is 0 and the power factor is 0.

$$S_{LA}=(.122\angle9.6^0)(.122\angle-9.6^0)100=1.49\angle0^0 \text{ W}$$
$$S_{LB}=(.122\angle110.4^0)(.122\angle-110.4^0)100=1.49\angle0^0 \text{ W}$$
$$S_{LC}=(.156\angle-120^0)(.156\angle120^0)50=1.22\angle0^0 \text{ W}$$
$$S_L=1.49\angle0^0+1.49\angle0^{00}+1.22\angle0^0 \text{ W}=4.2\angle0^0 \text{ W}$$

Wye Source to Delta Load

Figure 4-5 below shows a wye source connected to a delta load. This is a simplified diagram. **Zx** represents line impedance. Source impedances are assumed to be negligible. Voltages are measured with respect to the ground (the common of the sources).

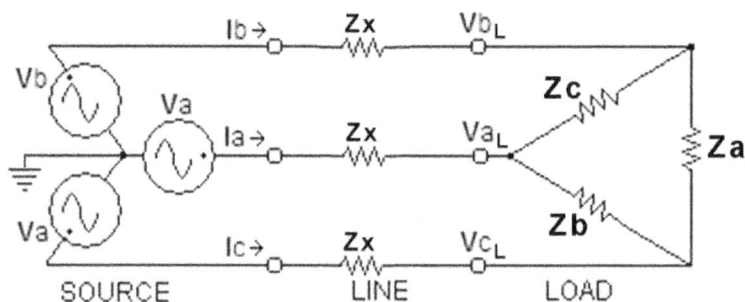

Figure 4-5

Equations for a wye-delta circuit:

The mesh current method is used in this example. The three current loops, **I1, I2,** and **I3** are used to write the mesh current equations. **I1** is the current through load **Zc**, **I2** is the current through load **Zb**, and **I3** is the current through load **Za**.

The values of **I1, I2,** and **I3** are used to calculate the source and line currents, **Ia, Ib,** and **Ic**. Refer to figure 4-6 below.

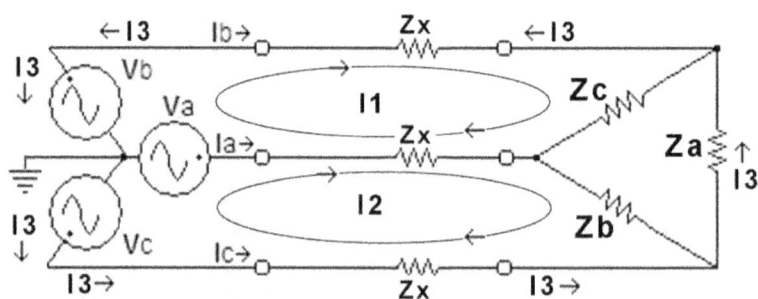

Figure 4-6

Loop Equations:

$$I_1 \text{Loop}: \quad I_1(2Zx+Zc) - I_2(Zx) - I_3(Zx) = Vb - Va$$
$$I_2 \text{ Loop}: \quad -I_1(Zx) + I_2(2Zx+Zb) - I_3(Zx) = Va - Vc$$
$$I_3 \text{ Loop}: \quad -I_1(Zx) - I_2(Zx) + I_3(2Zx+Za) = Vc - Vb$$

Given: $Ra = Rb = 100$. $Rc = 50$. $Rx = 20$. $Va = 12\angle0^0$

$$Va = 12\angle0^0, \quad Vb = 12\angle120^0, \quad Vc = 12\angle240^0$$

$I1 = .208\angle{-30} \quad I2 = .124\angle{-144.8} \quad I3 = .124\angle84.8$

$Ia = I2 - I1 = (.124\angle{-144.8}) - (.208\angle{-30}) = 0.283\angle{-6.6^0}$

$Ib = I1 - I3 = (.208\angle{-30}) - (.124\angle84.8) = 0.283\angle126.6^0$

$Ic = I3 - I2 = (.124\angle84.8) - (.124\angle{-144.8}) = 0.225\angle{-120^0}$

The total Power supplied by the source can be calculated independently for each phase.

$$S = Va(Ia^*) + Vb(Ib^*) + Vc(Ic^*)$$
$$S = (12\angle0^0)(.283\angle6.6^0) + (12\angle120^0)(.283\angle{-126.6^0}) + (12\angle{-120})(.225\angle120^0)$$
$$S = 3.396\angle6.6^0 + 3.396\angle{-6.6^0} + 2.7\angle0^0 = 9.447\angle0^0 \text{ Watts.}$$

The voltage at each node of the load can be calculated with ohm's law. These voltages will be used to calculate the line to line voltages for the load.

$$Va_L = Va - (Ia)Zx = 12 - (.283\angle{-6.6^0})20 = 6.41\angle5.82^0$$
$$Vb_L = Vb - (Ib)Zx = 12\angle120 - (.283\angle126.6^0)20 = 6.41\angle114.18^0$$
$$Vc_L = Vc - (Ic)Zc = 12\angle240 - (.225\angle{-120^0})20 = 7.5\angle{-120^0}$$

The line to line voltage across each load is calculated below.

$$Vab_L = Va_L - Vb_L = (6.41\angle5.82^0) - (6.41\angle114.18^0) = 10.4\angle{-30^0}$$
$$Vbc_L = Vb_L - Vc_L = (6.41\angle114.18^0) - (7.5\angle{-120^0}) = 12.39\angle84.8^0$$
$$Vca_L = Vc_L - Va_L = (7.5\angle{-120^0}) - (6.41\angle5.82^0) = 13.59\angle{-108.6}$$

The total power supplied to the load is the sum of the powers supplied to each phase.

$$S_L = Sab_L + Sbc_L + Sca_L = \frac{Vab_L(Zab_L^*)}{Zc} + \frac{Vbc_L(Zbc_L^*)}{Za} + \frac{Vca_L(Zca_L^*)}{Zb}$$

$$S_L = 2.163 + 1.535 + 1.847 = 5.54W$$

In rectangular format: $S_L = P + jQ = 5.54 + i0$, $P = 5.54$ and $Q = 0$.

Power Transfer Efficiency

$$P_T = \frac{P_L}{P} 100\% = \frac{5.54}{9.45} 100\% = 58.6\%$$

Balanced three wire wye-delta circuit:

Refer to the diagram on the right. A balanced load has equal load impedances and line voltages. The magnitudes of the phase currents, **Iza**, **Izb**, and **Izc** are equal and the magnitudes of the line currents, **Ia**, **Ib**, and **Ic** are equal.

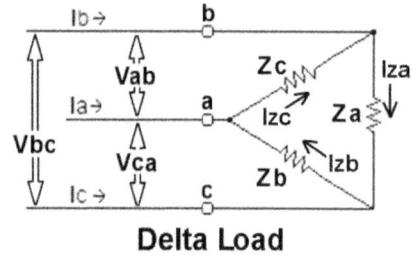

Delta Load

$$|Vab| = |Vbc| = |Vca| = |V_{LL}| \;\Rightarrow\; |Ia| = |Ib| = |Ic| = |I| \;\Rightarrow\; |Iza| = |Izb| = |Izc| = |I_P|$$

Set the reference phase of the line to line voltage **Vab** to $0°$ so that:

$$Vab = |Vab| \angle 0°, \qquad Vbc = |Vbc| \angle 120°, \qquad Vca = |Vca| \angle -120°.$$

$$Iza = \frac{|Vbc| \angle 120°}{|Za| \angle \theta°} = |I_P| \angle (120° - \theta°)$$

$$Izb = \frac{|Vca| \angle -120°}{|Zb| \angle \theta°} = |I_P| \angle (-120° - \theta°)$$

$$Izc = \frac{|Vab| \angle 0°}{|Zc| \angle \theta°} = |I_P| \angle (0° - \theta°)$$

Total power delivered to the load is:

$$S = Vab(Izc*) + Vbc(Iza*) + Vca(Izb*)$$

$$S = |V_{LL}||I_P|\angle\theta^0 + |V_{LL}||I_P|\angle\theta^0 + |V_{LL}||I_P|\angle\theta^0 = 3|V_{LL}||I_P|\angle\theta^0$$

$$S = 3|V_{LL}||I_P|\angle\theta^0 \qquad P = 3|V_{LL}||I_P|\cos\theta^0$$

It is desirable to have the power equation in terms of the line current rather than the phase current. Consider:

$$Ia = |Izc|\angle 0^0 - |Izb|\angle -120 = |I_P|\angle 0^0 - |I_P|\angle -120^0 = \sqrt{3}|I_P|\angle 30^0$$

Therefore : $|I| = \sqrt{3}|I_P|$

$$S = 3|V_{LL}|\frac{|I|}{\sqrt{3}}\angle\theta^0 = \sqrt{3}\ |V_{LL}||I|\angle\theta^0 \qquad P = 3|V_{LL}||I_P|\cos\theta^0$$

LTspice Simulation of the Wye to Delta Circuit

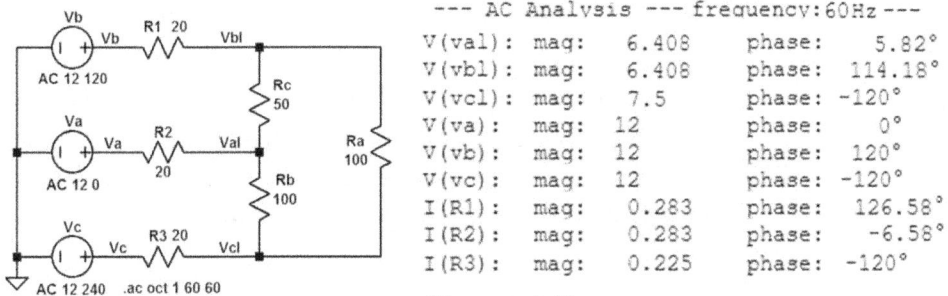

```
--- AC Analysis --- frequency:60Hz ---
V(val):   mag:   6.408    phase:     5.82°
V(vbl):   mag:   6.408    phase:   114.18°
V(vcl):   mag:   7.5      phase:  -120°
V(va):    mag:   12       phase:     0°
V(vb):    mag:   12       phase:   120°
V(vc):    mag:   12       phase:  -120°
I(R1):    mag:   0.283    phase:   126.58°
I(R2):    mag:   0.283    phase:    -6.58°
I(R3):    mag:   0.225    phase:  -120°
```

Figure 4-7

Note that the simulated currents I(R1), I(R2), and I(R3) are exactly equal to the calculated currents, **Ib, Ia**, and **Ic** for figure 4-6.

Delta Source to Wye Load

Figure 4-8 below shows a delta source connected to a wye load. This is a simplified diagram. **Zx** represents the line impedance. The source impedances are assumed to be negligible. Note the polarity of the sources.

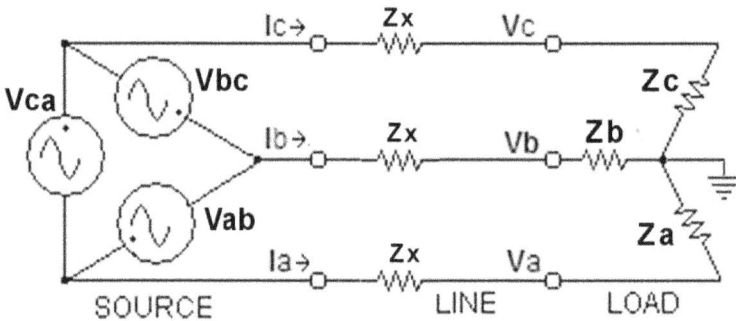

Figure 4-8

Equations for a delta-wye circuit:

The node voltage method is used in this example. The ground reference in figure 4-9 below is placed at the junction of **Vca** and **Vab**. The voltage, **V**, at the common node of the load and the currents, **Ia, Ib,** and **Ic,** are calculated. Note that the location of the ground node does not effect the line currents. Ohm's law is used to calculate **Va, Vb,** and **Vc.**

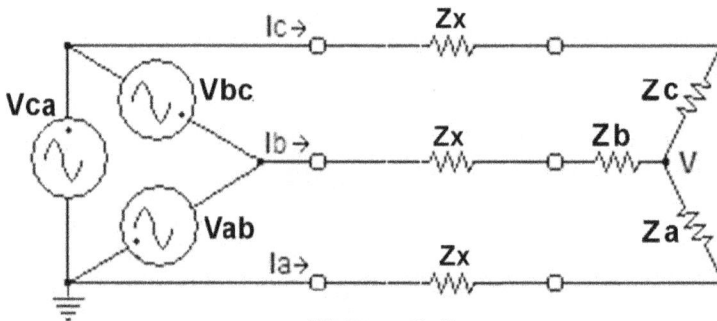

Figure 4-9

Given: Zb = Zc = 100,   Za = 30 + j30.   Zx =10,

Vab = $20\angle 0^0$ Vrms,   Vbc = $20\angle 120^0$ Vrms,   Vca = $20\angle -120^0$ Vrms.

Node voltage equation:

$$\frac{\mathbf{Vca}-\mathbf{V}}{\mathbf{Zx}+\mathbf{Zc}}+\frac{\mathbf{Vab}-\mathbf{V}}{\mathbf{Zx}+\mathbf{Zb}}+\frac{0-\mathbf{V}}{\mathbf{Zx}+\mathbf{Za}}=\frac{-10-j17.32-\mathbf{V}}{110}+\frac{-20-\mathbf{V}}{110}+\frac{0-\mathbf{V}}{40+j30}=0$$

$\mathbf{V} = (\text{-}5.669\text{-}j6.592) = 8.693\angle\text{-}130.7$

Note that the voltage **Vbc** was not needed in this calculation. It is a property of a delta source that it will continue to supply 3-phase power even if one of the sources becomes an open circuit. Of course with one of the sources missing the supplies power capability will be reduced.

Line currents:

$$\mathbf{Ic}=\frac{-10-j17.32-(-5.669-j6.592)}{110}=0.1052\angle-111.98^0$$

$$\mathbf{Ib}=\frac{-20-(-5.669-j6.592)}{110}=0.1434\angle155.3^0$$

$$\mathbf{Ia}=\frac{0-(-5.669-j6.592)}{40+j30}=0.1739\angle112.43^0$$

Load voltages:

The following voltages are calculated with respect to the ground at the junction of the load resistors.

Val = Ia Za = $(0.1739\angle112.43^0)(30 + j30)$ = $7.38\angle57.28$

Vbl = IbRb = $(0.1434\angle155.3^0)(30 + j30)$ = $14.34\angle155.3^0$.

Vcl = IcRc = $(0.1052\angle\text{-}111.98^0)(30 + j30)$ = $10.52\angle\text{-}111.98^0$

Simulating a delta source

The phases of the delta source are effectively connected in series. A circulating current may occur if the magnitudes and phase angles of the source are not properly specified for each phase. Simulation programs will require a resistance in series with each source. In LTspice, the source resistance may be specified.

LTspice Simulation of the Delta to Wye Circuit

The series resistance was set to 0.1 for each phase by right-clicking on each source. The results below closely agree with the calculations.

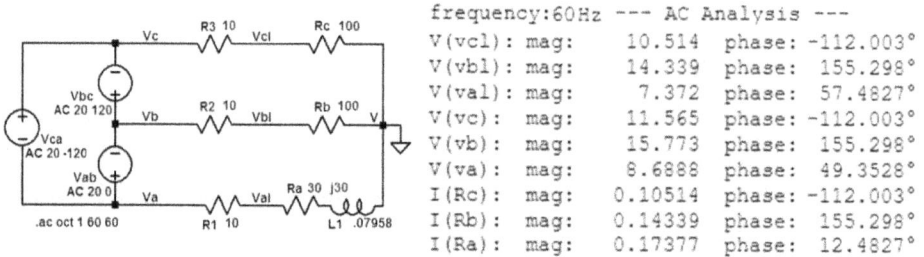

```
frequency:60Hz --- AC Analysis ---
V(vcl): mag:    10.514  phase: -112.003°
V(vbl): mag:    14.339  phase:  155.298°
V(val): mag:     7.372  phase:   57.4827°
V(vc):  mag:    11.565  phase: -112.003°
V(vb):  mag:    15.773  phase:  155.298°
V(va):  mag:     8.6888 phase:   49.3528°
I(Rc):  mag:     0.10514 phase: -112.003°
I(Rb):  mag:     0.14339 phase:  155.298°
I(Ra):  mag:     0.17377 phase:   12.4827°
```

Figure 4-10

Wye Equivalent of a Delta Source

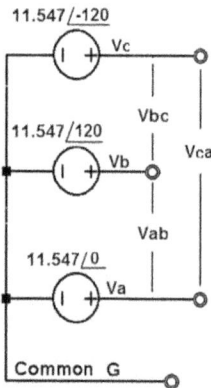

Figure 4-11

A wye connected source can supply the same line to line voltages as a delta source. Refer to figure 4-11.

The magnitude of the wye equivalent source voltages must be equal to the magnitude of the delta source voltages divided by the square root of 3.

$$|\mathbf{V}|_{LINE-NEUTRAL} = \frac{|\mathbf{V}|_{LINE-LINE}}{\sqrt{3}}$$

The line-to-line phase angles are shifted by 30 compared to the line-to-neutral phase angles.

Example calculation:

A 20 volt per phase delta source is obtained by setting the line to neutral voltages to 11.547 volts.

$$\mathbf{Vab} = \mathbf{Va} - \mathbf{Vb} = (11.547\angle0^0) - (11.547\angle120^0) = 20\angle30^0$$

$$\mathbf{Vbc} = \mathbf{Vb} - \mathbf{Vc} = (11.547\angle120^0) - (11.547\angle-120^0) = 20\angle90^0$$

$$\mathbf{Vca} = \mathbf{Vc} - \mathbf{Va} = (11.547\angle-120^0) - (11.547\angle0^0) = 20\angle-150^0$$

LTspice Simulation using the Wye Equivalent Source

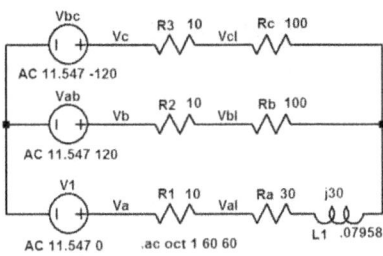

```
frequency:60Hz  --- AC Analysis ---
V(vcl):  mag:   10.517   phase: -142.013°
V(vbl):  mag:   14.345   phase:  125.297°
V(val):  mag:    7.3763  phase:   27.474°
V(vc) :  mag:   11.569   phase: -142.013°
V(vb) :  mag:   15.780   phase:  125.297°
V(va) :  mag:    8.6930  phase:   19.3442°
I(Rc):   mag: 0.10517    phase: -142.013°
I(Rb):   mag: 0.14345    phase:  125.297°
I(Ra):   mag: 0.17385    phase:  -17.5259°
```

Figure 4-12

Compare the results shown in figure 4-12 above with the results shown in figure 4-10. All of the magnitudes are the same and all of the phase angles are 30 degrees less.

PSpice Simulation

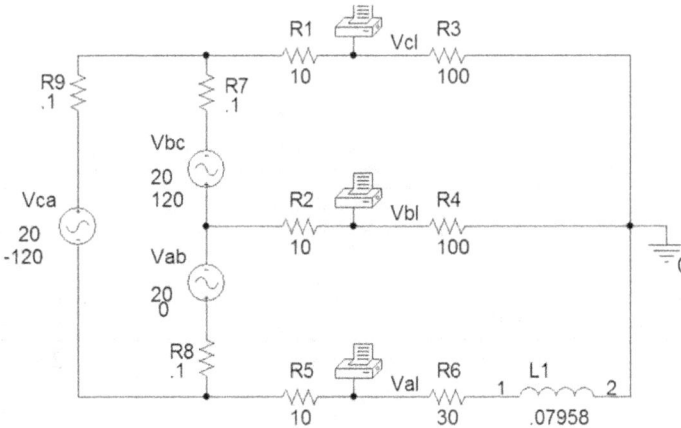

Figure 4-13

Results

VM(VAL)	7.373E+00	VP(VAL)	5.748E+01
VM(VCL)	1.051E+01	VP(VCL)	-1.120E+02
VM(VBL)	1.434E+01	VP(VBL)	1.553E+02

71

Delta Source to Delta Load

Figure 4-14 below shows a delta source connected to a delta load. This is a simplified diagram. **Zx** represents the line impedances. The source impedances are assumed to be negligible.

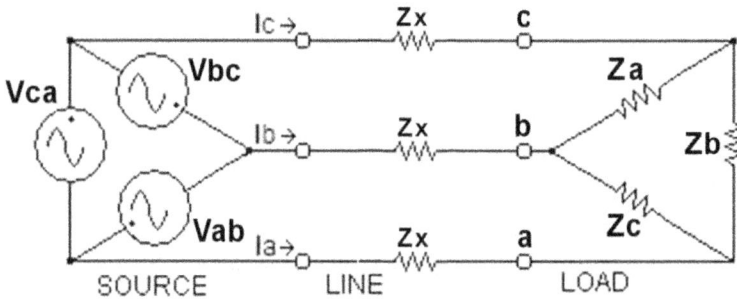

Figure 4-14

Equations for a delta-wye circuit:

The mesh method is used to in this example.

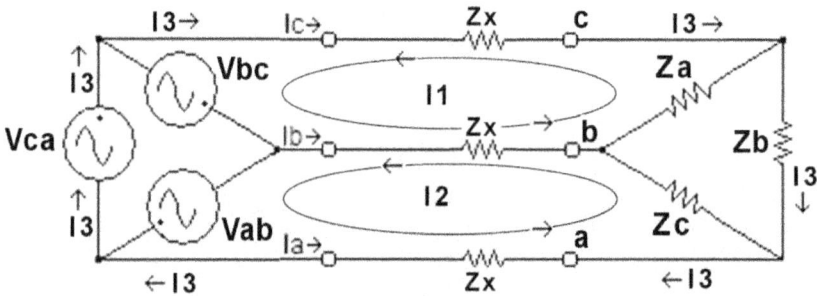

Figure 4-15

I_1 Loop: $\quad I_1(2Zx+Za) - I_2(Zx) - I_3(Zx)=Vbc$

I_2 Loop: $\quad -I_1(Zx)+ I_2(2Zx+Zc) -I_3(Zx)=Vab$

I_3 Loop: $\quad -I_1(Zx) - I_2(Zx) + I_3(2Zx+Zb)=Vca$

Given: Zb = Zc = 100, Za = 30 + j30. Zx =10, Vab = 20∠0⁰,

Vbc = 20∠120⁰, Vca = 20∠-120⁰.

$\mathbf{I}_1$ Loop: $\quad \mathbf{I}_1(50+j30) - \mathbf{I}_2(10) - \mathbf{I}_3(10) = 20\angle 120^0$

$\mathbf{I}_2$ Loop: $\quad -\mathbf{I}_1(10) + \mathbf{I}_2(120) - \mathbf{I}_3(10) = 20\angle 0^0$

$\mathbf{I}_3$ Loop: $\quad -\mathbf{I}_1(10) - \mathbf{I}_2(10) + \mathbf{I}_3(120) = 20\angle -120^0$

$I_1 = 0.01066 + j0.32015 = 0.3203\angle 88.09^0$
$I_2 = 0.16180 + j0.01699 = 0.1627\angle 6^0$
$I_3 = -0.06896 - j0.11623 = 0.1352\angle\text{-}120.68^0$

$Ia = I2 - I3 = 0.2665\angle 30^0$
$Ib = I1 - I2 = 0.3388\angle 116.5^0$
$Ic = I3 - I1 = 0.4436\angle\text{-}100.34^0$

$\mathbf{Vab_L = I_2Zc} \qquad \mathbf{Vbc_L = I_1Za} \qquad \mathbf{Vca_L = I_3Zb}$

$\mathbf{Vab_L = I_2Zc =} (0.1627\angle 6^0)(100) = 16.27\angle 6^0$
$\mathbf{Vbc_L = I_1Za =} (0.3203\angle 88.09^0)(30 + j30) = 13.59\angle 133.09^0$
$\mathbf{Vca_L = I_3Zb =} (0.1352\angle\text{-}120.68^0)(100) = 13.52\angle\text{-}120.68^0$

Experiment 4a: Wye to Wye Circuit

This experiment involves the measurement of voltage amplitudes and phase angles in a 3-phase wye-wye circuit. It demonstrates the effects of balanced and un-balanced loads in four wire wye and three wire wye circuits.

All measurements are from line to neutral. Line currents and line to line voltages are calculated from the line to neutral measurements. Line resistance is represented by 10Ω resistors

Equipment and Parts

Voltage Source: 3-Phase, 12Vp-p , Oscilloscope, and Breadboard.
Resistors: four 10Ω, four 390Ω, all ¼ watt, 5%.

Procedure: Part 1, Balanced Load

Connect the circuit in figure 4-16 below. The 3-phase source outputs are set to: **Va** = 12∠0⁰ V p-p, **Vb** = 12∠120⁰ V p-p, **Vc** = 12∠-120⁰ V p-p.

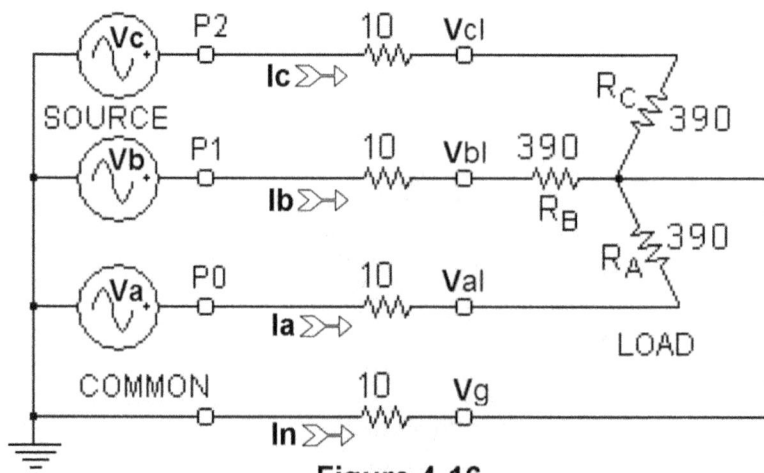

Figure 4-16

2. Connect channel 1 of the oscilloscope to P0. Trigger on channel 1. Connect channel 2 of the oscilloscope to P1. Measure and record the magnitude of **Va** and the magnitude and phase angle of **Vb**.

Va:_____ V p-p   θa: 0⁰   **Vb:**_____ V p-p   θb:_____ ⁰

74

3. Connect channel 2 of the oscilloscope to P2. Measure and record the magnitude of **Vc** and the magnitude and phase angle of **Vc**.

Vc:_____ V p-p   **θc:**_____ 0

4. **Four wire circuit:** Measure and record the voltages and phase angles of **Vg, Val, Vbl,** and **Vcl** with channel 2 of the oscilloscope. Record below:

node	Vg	Val	Vbl	Vcl
Mag. V p-p				
Angle Deg.				

5. **Three wire circuit:** Open the neutral line by removing the 10-ohm resistor between the common and **Vg**.

6. Measure and record the voltages and phase angles of **Vg, Val, Vbl,** and **Vcl** with channel 2 of the oscilloscope. Record below:

Step 6 node	Vg	Val	Vbl	Vcl
Mag. V p-p				
Angle Deg.				

Part 2: Unbalanced Load

1. Connect a 390 ohm resistor in parallel with RC so that the net resistance is 195 ohms. Reconnect the 10Ω resistor between the COMMON and node **Vg**.

2. **Four wire circuit:** Measure the magnitude and phase angle of the voltages at nodes **Vg, Val. Vbl,** and **Vcl** with the oscilloscope channel 2 and record results below.

Step 2 node	Vg	Val	Vbl	Vcl
Magnitude p-p				
Phase, Deg.				

3. Open the neutral line by removing the 10Ω resistor between the COMMON and node **Vg**.

4. **Three wire circuit:** Measure the magnitude and phase angle of the voltages at nodes **Vg, Val, Vbl,** and **Vcl** with the oscilloscope channel 2 and record results below.

Step 4 node	Vg	Val	Vbl	Vcl
Magnitude p-p				
Phase, Deg.				

Analysis Part 1: Balanced Load

1. Calculate the theoretical current per phase for the four wire circuit. Calculate the measured current per phase using the voltage measurements in step 4 of the procedure. Compare the calculated and measured results. Compare the magnitudes of the line currents to the neutral current.

2. Calculate the theoretical value of **Vg** for the three wire circuit using the node voltage method. Calculate the phase currents. Compare the results to procedure step 6 measurements.

3. Calculate the power supplied to the balanced four wire load. Calculate the power loss in the transmission line. Calculate the percent   efficiency of the circuit.

Analysis Part 2: Unbalanced Load

1. Calculate the theoretical current per phase for the four wire unbalanced circuit.   Calculate the measured current per phase using the voltage measurements in step 2 of the procedure. Compare the calculated and measured results. Compare the magnitudes of the line currents to the neutral current.

2. Calculate the theoretical value of **Vg** for the three wire unbalanced circuit (use node voltage method).   Calculate the phase currents. Compare the results to procedure step 5 measurements.

3. Simulate the unbalanced circuit with and without the neutral connection.   Refer to the example in the next section.

4. Use Kirchhoff's current law to show that the neutral current in the four wire circuit is the result of the unbalance of the phase currents.

LTspice Example: Unbalanced Three Wire Wye

AC analysis is performed using three AC sources as shown in figure 4-17 below. Right click on each source to set its AC amplitude and phase angle. In this example the amplitude of each source is considered to be in volts peak-to-peak. Therefore the units for the simulation results will also be in volts peak-to-peak.

Simulation was set to AC Analysis, octave sweep, one point per octave, from 60Hz to 60Hz (find the command line on the schematic in figure 4-17 below).

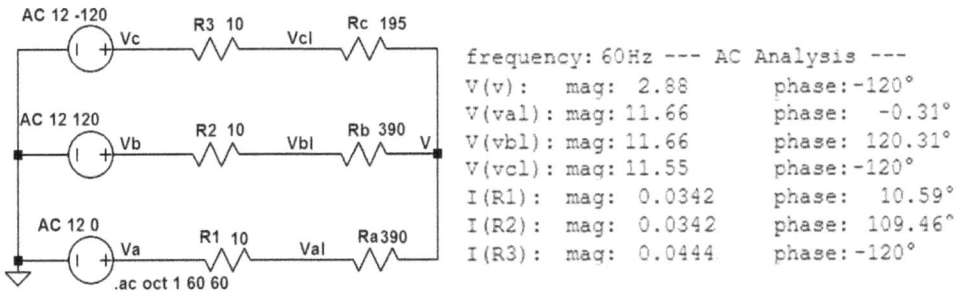

```
AC 12 -120
        Vc      R3 10      Vcl      Rc 195
                                            frequency: 60Hz --- AC Analysis ---
                                            V(v):    mag:  2.88    phase:-120°
AC 12 120                                   V(val): mag: 11.66    phase:   -0.31°
        Vb      R2 10      Vbl      Rb 390  V(vbl): mag: 11.66    phase: 120.31°
                                         V  V(vcl): mag: 11.55    phase:-120°
                                            I(R1):   mag: 0.0342   phase:  10.59°
AC 12 0                                     I(R2):   mag: 0.0342   phase: 109.46°
        Va      R1 10      Val      Ra390   I(R3):   mag: 0.0444   phase: -120°
              .ac oct 1 60 60
```

Figure 4-17

The actual voltage across each load resistance is calculated below:

V_{RA} = **Val** – **V** = $11.66\angle-0.31^0 - 2.88\angle-120^0 = 13.32\angle10.51^0$

V_{RB} = **Vbl** – **V** = $11.66\angle120.31^0 - 2.88\angle-120^0 = 13.32\angle109.48^0$

V_{RC} = **Vcl** – **V** = $11.55\angle-120^0 - 2.88\angle-120^0 = 8.67\angle-120^0$

PSpice: Balanced and Unbalanced Three Wire Wye

This example does not use the part values in the experiment. Enter your own part values. Three phase circuits are simulated using three "VAC" sources from the source library.

Double click on each source to open its property editor to set its phase angle. Click on "Display" and set each to display its name and value.

The printers "VPRINT1" are in the "SPECIAL" library. Double click each one to open their property editor. Enable them to read AC magnitude and phase by entering "ok" in the AC, Magnitude, and Phase columns.

Use "AC Sweep analysis" with the start and end frequencies set to 60 Hertz and the number of points set to 1.

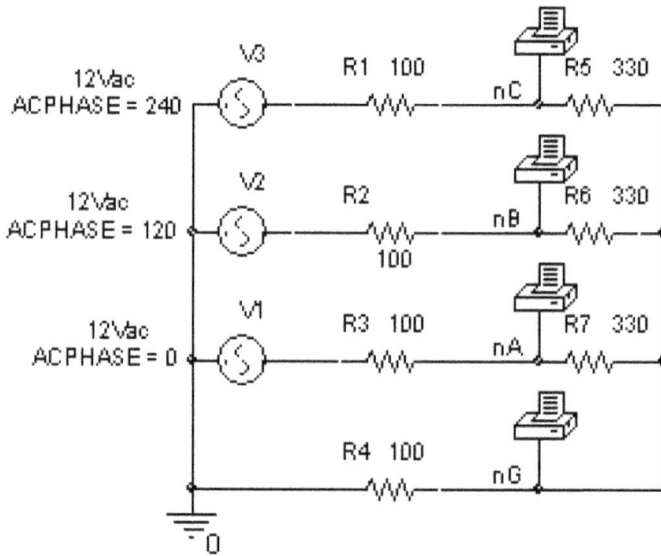

Figure 4-18

Refer to the schematic diagram in figure 4-18 above and the edited simulation results below for a balanced and unbalanced load. The last two columns show the results when the load is unbalanced by replacing one of the 330 Ω resistors with a 1000 Ω resistor.

60 Hertz	R5=330, R6=330, R7=330.		R5=1K, R6=330, R7=330.	
Node	Mag V p-p	Phase Deg.	Mag V p-p	Phase Deg.
nC	9.209	-120	10.081	-120
nB	9.209	120	9.339	118.7
nA	9.209	0	9.339	1.35
nG	0	N.A.	1.092	60.0

Experiment 4b: Three-Phase Power / Delta Connection

This experiment involves the measurement of voltage amplitudes and phase angles in a Three-phase wye-delta circuit. It demonstrates the effects of balanced, unbalanced, and reactive delta connected loads.

All measurements are from line to neutral. Line current and line to line voltage is calculated from the line to neutral measurements. Line resistance is represented by 10Ω resistors

Equipment and Parts

Voltage Source: Three Phase, 12Vp-p , Oscilloscope, and Breadboard.
Resistors: Three 10Ω, four 1k, all ¼ watt, 5%. Capacitor: 4.7µF ceramic.

Procedure: Part 1, Balanced Load

Connect the circuit in figure 4-19 below. The 3-phase source outputs are set to: **Va** = 12∠0⁰ V p-p, **Vb** = 12∠120⁰ V p-p, **Vc** = 12∠-120⁰ V p-p.

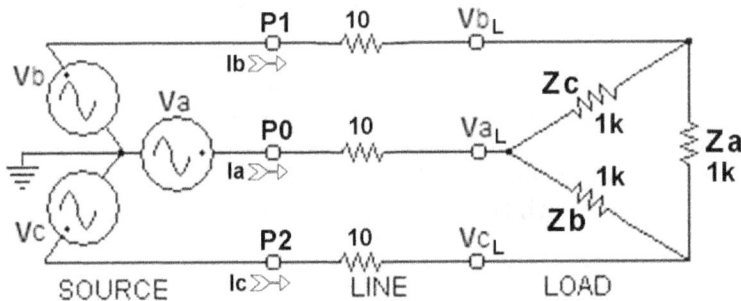

Figure 4-19

2. Connect channel 1 of the oscilloscope to P0. Trigger on channel 1. Connect channel 2 of the oscilloscope to P1. Measure and record the magnitude of **Va** and the magnitude and phase angle of **Vb**.

Va:_____ V p-p **θa:** _0⁰_   **Vb:**_____ V p-p **θb:**_____⁰

3. Connect channel 2 of the oscilloscope to P2. Measure and record the magnitude of **Vc** and the magnitude and phase angle of **Vc**.

Vc:_____ V p-p **θc:**_____⁰

4. **Balanced Load:** Measure and record the voltages and phase angles of **Val**, **Vbl**, and **Vcl** with channel 2 of the oscilloscope. Record below:

Step 4 node	Val	Vbl	Vcl
Mag. V p-p			
Angle Deg.			

5. **Unbalanced Load:** Connect a 1k resistor in parallel with Rc.

6. Measure and record the voltages and phase angles of **Val**, **Vbl**, and **Vcl** with channel 2 of the oscilloscope. Record below:

Step 6 node	Val	Vbl	Vcl
Mag. V p-p			
Angle Deg.			

7. **Reactive Load:** Remove the 1k resistor in parallel with Rc. Connect a 4.7µF (non-polarized ceramic) in parallel with Rc.

8. Measure the magnitudes and phase angles of the voltages at nodes **Val. Vbl**, and **Vcl** with the oscilloscope channel 2 and record results below.

Step 2 node	Val	Vbl	Vcl
Mag. V p-p			
Phase, Deg.			

Analysis

1. Use the mesh current method to calculate the theoretical line current per phase and load current per phase for the balanced load. Calculate the measured line currents using measurements in step 4 of the procedure. Compare the calculated and measured results.

2. Use the mesh current method to calculate the theoretical line current per phase and load current per phase for the unbalanced load. Calculate the measured line currents using measurements in step 6 of the procedure. Compare the calculated and measured results.

3. Use the mesh current method to calculate the theoretical line current per phase and load current per phase for the reactive load. Calculate the measured line currents using the measurements in step 8 of the procedure. Compare the calculated and measured results.

4. Simulate the reactive load circuit and compare the results to your measurements.

5. Calculate the total average and reactive power delivered by the source and supplied to the load.

6. Calculate the circuit's power factor and efficiency.

LTspice Example

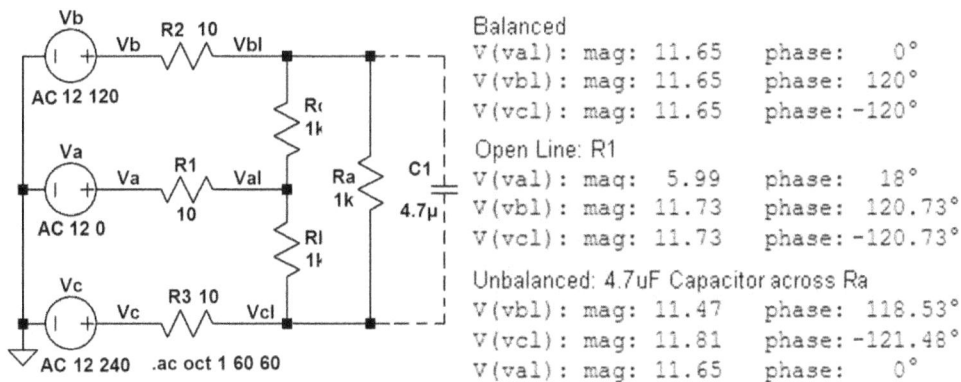

```
Balanced
V(val):  mag:  11.65    phase:     0°
V(vbl):  mag:  11.65    phase:   120°
V(vcl):  mag:  11.65    phase:  -120°

Open Line: R1
V(val):  mag:   5.99    phase:    18°
V(vbl):  mag:  11.73    phase:   120.73°
V(vcl):  mag:  11.73    phase:  -120.73°

Unbalanced: 4.7uF Capacitor across Ra
V(vbl):  mag:  11.47    phase:   118.53°
V(vcl):  mag:  11.81    phase:  -121.48°
V(val):  mag:  11.65    phase:     0°
```

Figure 4-20

The circuit in figure 4-20 above was simulated for a balanced load and reactive load. In addition, a simulation was done with resistor R1 open (simulates an open transmission line).

TI-89 Example

Let $V(nA) = x$, $V(nB) = y$, $V(nC) = z$.

Equations:

$$\frac{x - 12\angle 0}{100 + 37.7i} + \frac{x - y}{330} + \frac{x - z}{330} = 0 \text{ and } \frac{y - 12\angle 120}{100 + 37.7i} + \frac{y - x}{330} + \frac{y - z}{330} = 0$$

$$\text{and } \frac{z - 12\angle 240}{100 + 37.7i} + \frac{z - x}{330} + \frac{z - y}{330} = 0$$

TI-89 input:

Balanced case:

csolve((x-12)/(100+37.7i)+(x-y)/330+(x-z)/330=0 and
(y-(12∠120))/(100+37.7i)+(y-x)/330+(y-z)/330=0 and
(z-(12∠240))/(100+37.7i)+(z-x)/330+(z-y)/330=0,{x,y,z})

x = (6.187∠-10.8) y = (6.187∠109.8)z = (6.187∠-130.2)
Unbalanced case:

csolve((x-12)/(100+37.7i)+(x-y)/330+(x-z)/220=0 and
(y-(12∠120))/(100+37.7i)+(y-x)/330+(y-z)/330=0 and
(z-(12∠240))/(100+37.7i)+(z-x)/220+(z-y)/330=0,{x,y,z})

x = (5.611∠-15.12)   y = (6.187∠109.82)   z = (5.477∠-127.29)

Ch 5: Three-Phase Transformers

Three-phase transformers are used for voltage step up and step down and for delta to wye and wye to delta conversion. A bank of three single-phase transformers may be used, however more economical single-core transformers are available. Figure 5-1 below illustrates the configuration of the windings on a single-core three-phase transformer.

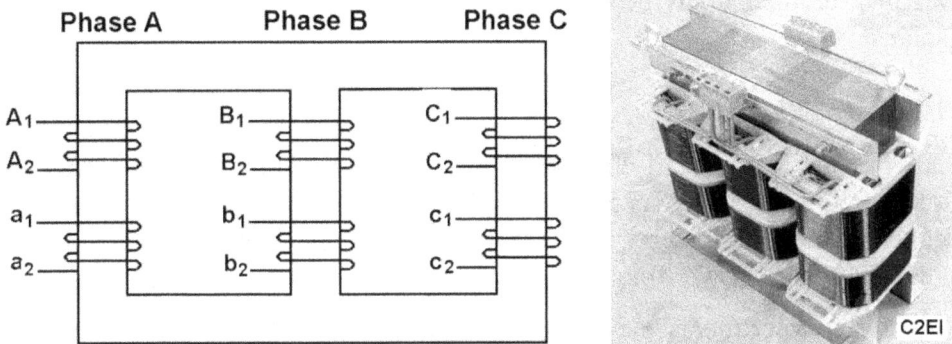

Figure 5-1

The transformer windings are typically configured as delta to delta, delta to wye, or wye to delta. There is a configuration of the delta to delta that can be implemented with just two single-phase transformers as shown below in figure 5-2. This is called the "open delta" and is used for some low power three-phase distributions.

If the primary voltages V_{AB} and V_{BC} are given, the voltage V_{CA} can be calculated, as shown below. Given the turns ratio of the transformer, the magnitudes of the secondary voltages can be calculated. The corresponding angles of the secondary voltages will be the same as the primary (note the phasing dots on the transformer).

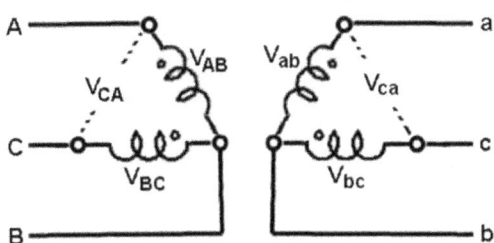

Given:

$V_{AB} = 208\angle0^0$, $V_{BC} = 208\angle120^0$

Then:

$V_{CA} + V_{BC} + V_{AB} = 0$

$V_{CA} = -(V_{BC} + V_{AB})$

$V_{CA} = -208\angle60^0 = 208\angle-120^0$

Figure 5-2

Three-Phase Transformer Experiments

The experiments in this chapter are designed to use a low voltage 3-phase voltage source. This could be derived from a 3-phase outlet using a 3-phase step-down transformer, or from a 1-phase to 3-phase converter circuit. Of course, the experiments can be modified to use other voltages and transformers.

All oscilloscope measurements are made with respect to a common ground. Line to line voltages may be calculated from line to neutral voltages. Line to line measurements may be made if the measuring instrument is known to be isolated.

Using 400 Hertz

Experiments using small 200mW to 400mW transformers do not work well at low frequencies such as 50Hz or 60Hz. However, larger transformers or a higher frequencies may be used. Transformer experiments in this chapter are done at 400Hz. Lower frequencies may be used with larger transformers.

All of the experiments may be performed using a 12V p-p, line to neutral source. A simple "Phase Tripler" circuit is described in the appendix which may be built on a breadboard. The Phase Tripler may be set for 60Hz or 400Hz depending on the values of two of its resistors and capacitors.

A 400Hz source has the advantage that smaller inductances and capacitances may be used when doing 3-phase breadboard experiments. Practical breadboard experiments, especially when using ¼ watt resistors, involve load impedances between 300 and 1000 ohms.

Because of physical size and cost, inductance should be less than 68mH and capacitance less than 0.5µF. Relatively low cost, 1µF to 10µF ceramic capacitors may also be used. Reactances are given below for comparison:

Value	Reactance at 60Hz	Reactance at 400Hz
68mH	25.64Ω	170.90Ω
0.47µF	5644Ω	847Ω

Experiment 5a: Wye-Delta transformer

The wye-delta configuration is typically used to step up the voltages of a wye configured generator. For example, a hydroelectric generator may produce 18KV per phase (line to neutral). A wye-delta transformer steps up the voltage to over 500KV to feed many miles transmission lines. In this experiment the delta side of the transformer is "corner grounded".

Equipment and Parts

Function Generator, Oscilloscope, DMM, and Breadboard.
Three-Phase Source
Transformer, Three 500Ω CT to 500Ω CT, 400mW (see appendix 2).
    Recommended: ZICON 42TU500-RC (from Mouser Electronics)
Resistors: Three 10Ω, three 1k, all ¼ watt, 5%.

Note: The specified 400mW transformer may be used at 60Hz. The results will show a somewhat lower efficiency.

Procedure: Part 1, No Fault

1. Connect the circuit in Figure 5-3 below. **J1** is a jumper that will be connected for parts 1 and 2 of this experiment and disconnected for part 3. The frequency of the 3-phase source is 400Hz. The amplitude of each phase is 12V p-p at the phase angles indicated.

2. Connect oscilloscope channel 1 to **Va.** Trigger on channel 1. **Va** will be the reference phase for the entire experiment.

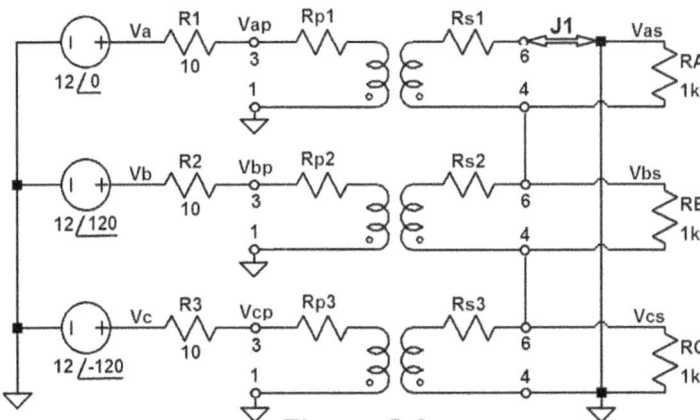

Figure 5-3

3. Measure and record the magnitude of **Va** and the magnitude and phase angle of **Vb**.

Va:_____ V p-p   **θa:** <u>0⁰</u>   **Vb:**_____ V p-p   **θb:**_____ ⁰

Connect channel 2 of the oscilloscope to P2. Measure and record the magnitude of **Vc** and the magnitude and phase angle of **Vc**.

Vc:_____ V p-p   **θc:**_____ ⁰

4. Measure and record the primary voltages **Vap**, **Vbp**, and **Vcp**.

node		Vap	Vbp	Vcp
Mag. V p-p				
Angle Deg.				

5. Measure and record the secondary voltages **Vas**, **Vbs**, and **Vcs**.

node		Vas	Vbs	Vcs
Mag. V p-p				
Angle Deg.				

Procedure: Part 2, Open Primary Winding

1. Remove the resistor R1. Leave the jumper **J1 connected**.

2. Measure and record the primary voltages **Vap**, **Vbp**, and **Vcp**.

node		Vap	Vbp	Vcp
Mag. V p-p				
Angle Deg.				

3. Measure and record the secondary voltages **Vas**, **Vbs**, and **Vcs**.

node		Vas	Vbs	Vcs
Mag. V p-p				
Angle Deg.				

Procedure: Part 3, Open Secondary Winding

1. Remove the jumper **J1**. Replace the resistor R1.
2. Measure and record the primary voltages **Vap**, **Vbp**, and **Vcp**.

node		Vap	Vbp	Vcp
Mag. V p-p				
Angle Deg.				

3. Measure and record the secondary voltages **Vas**, **Vbs**, and **Vcs**.

node		Vas	Vbs	Vcs
Mag. V p-p				
Angle Deg.				

Analysis, Part 1

Use the part 1 measurements to make the calculations below.

1. Calculate the primary currents, **Ia**, **Ib**, and **Ic**.
2. Calculate the total complex power, **S**, and total average power, P. Calculate the circuit's power factor.
3. Calculate the secondary line to line voltages and the total power, P, delivered to the load.
5. Calculate the efficiency of the circuit and of the 3-phase transformer.
6. Compare your results to a simulation.

Analysis, Part 2

Use the part 2 measurements to make the calculations below.

1. Calculate the primary currents, **Ia**, **Ib**, and **Ic**.

2. Calculate the total complex power, **S**, and total average power, P. Calculate the circuit's power factor.

3. Calculate the secondary line to line voltages and the total power, P, delivered to the load.

4. Compare your results to a simulation.

5. Explain the voltage across the load resistor, Ra, when the primary winding of the phase **a** is open (R1 removed).

Analysis, Part 3

Use the part 3 measurements to make the calculations below.

1. Calculate the primary currents, **Ia**, **Ib**, and **Ic**.

2. Calculate the total complex power, **S**, and total average power, P. Calculate the circuit's power factor.

3. Calculate the secondary line to line voltages and the total power, P, delivered to the load.

4. Compare your results to a simulation.

5. Explain the voltage across the load resistor, Ra, when the secondary of phase **a** is open (**J1** is removed).

LTspice Simulation

Simulate your circuit at the frequency that you used in the experiment using your measured transformer resistances. Refer to figure 5-4 below.

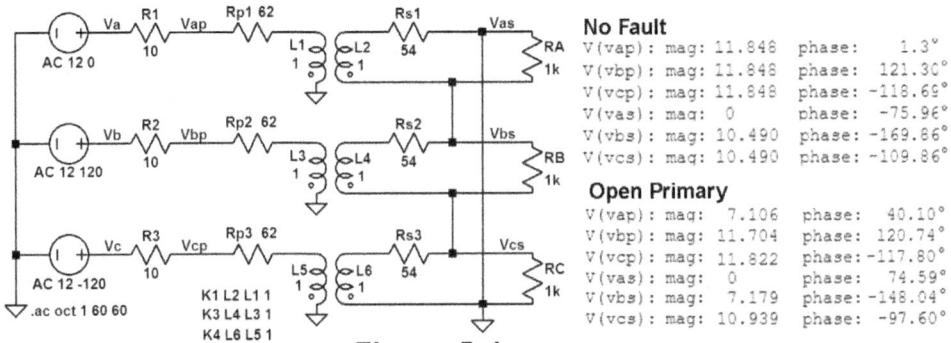

No Fault
```
V(vap):  mag: 11.848   phase:     1.3°
V(vbp):  mag: 11.848   phase:  121.30°
V(vcp):  mag: 11.848   phase: -118.69°
V(vas):  mag:  0        phase:  -75.96°
V(vbs):  mag: 10.490   phase: -169.86°
V(vcs):  mag: 10.490   phase: -109.86°
```

Open Primary
```
V(vap):  mag:  7.106   phase:   40.10°
V(vbp):  mag: 11.704   phase:  120.74°
V(vcp):  mag: 11.822   phase: -117.80°
V(vas):  mag:  0        phase:   74.59°
V(vbs):  mag:  7.179   phase: -148.04°
V(vcs):  mag: 10.939   phase:  -97.60°
```

Figure 5-4

The secondary line to line voltages are calculated below using the data from the simulation of the circuit in figure 5-4.

No Fault

$Vabs = Vas - Vbs = 0 - 10.49\angle -169.86 = 10.49\angle 10.14$

$Vbcs = Vbs - Vcs = 10.49\angle -169.86 - 10.49\angle -109.86 = 10.49\angle 130.14$

$Vcas = Vcs - Vas = 10.49\angle -109.86 - 0 = 10.49\angle -109.86$

Open Primary

$Vabs = Vas - Vbs = 0 - 7.18\angle -148.04 = 7.18\angle 31.96$

$Vbcs = Vbs - Vcs = 7.18\angle -148.04 - 10.94\angle -97.6 = 8.44\angle 123.4$

$Vcas = Vcs - Vas = 10.94\angle -97.67 - 0 = 10.94\angle -97.67$

Experiment 5b: Delta-Wye transformer

A delta-wye configuration is typically used to step down transmission line and distribution line voltages. For example, it may be used to step down a 13.2KV distribution line voltage to a 240V/120V single-phase residential supply or a 208V/120V Three-phase commercial supply. A four wire service can provide Three phases of 208V line to line voltage and three phases of 120V line to neutral voltage.

Equipment and Parts

Function Generator, Oscilloscope, DMM, and Breadboard.
Resistors: Three 10Ω, three 1k, all ¼ watt, 5%.
Capacitor: 0.68uF, 5%, Film type (4.7µF ceramic for 60Hz).
Transformer, Three 500Ω CT to 500Ω CT, 400mW (see appendix 2).
    Recommended: ZICON 42TU500-RC (from Mouser Electronics)

Note: The specified 400mW transformer may be used at 60Hz. If 60Hz is used, use a 4.7µF capacitor instead of the 0.68µF. The results will show a somewhat lower efficiency.

Procedure: Part 1, No Fault

1. Connect the circuit in Figure 5-5 below. **J1** is a jumper that will be connected for part 1 and part 3 of this experiment, and disconnected for part 2. The frequency of the Three-phase source is 400Hz. The amplitude of each phase is 12V p-p at the phase angles indicated.

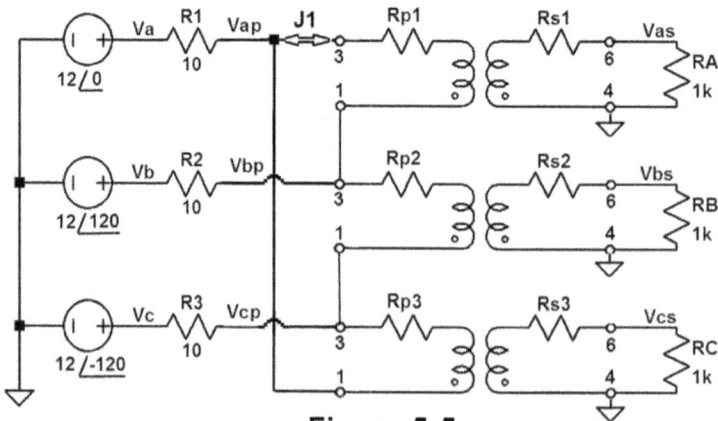

Figure 5-5

2. Connect oscilloscope channel 1 to **Va**. Trigger on channel 1. **Va** will be the reference phase for the entire experiment.

3. Measure and record the magnitude of **Va** and the magnitude and phase angle of **Vb**.

   Va:_____ V p-p   **θa:** 0⁰   **Vb:**_____ V p-p   **θb:**_____⁰

   Connect channel 2 of the oscilloscope to P2. Measure and record the magnitude of **Vc** and the magnitude and phase angle of **Vc**.

   Vc:_____ V p-p   **θc:**_____⁰

4. Measure and record the primary voltages **Vap**, **Vbp**, and **Vcp**.

node	Vap	Vbp	Vcp
Mag. V p-p			
Angle Deg.			

5. Measure and record the secondary voltages **Vas**, **Vbs**, and **Vcs**.

node	Vas	Vbs	Vcs
Mag. V p-p			
Angle Deg.			

Procedure: Part 2, Open Primary Winding

1. Remove the jumper **J1**.

2. Measure and record the primary voltages **Vap**, **Vbp**, and **Vcp**.

node	Vap	Vbp	Vcp
Mag. V p-p			
Angle Deg.			

91

3. Measure and record the secondary voltages **Vas**, **Vbs**, and **Vcs**.

node		Vas	Vbs	Vcs
Mag. V p-p				
Angle Deg.				

Procedure: Part 3, Reactive Load

1. Reconnect the jumper **J1**. Connect a 0.68μF capacitor across Rc (4.7μF for 60Hz) .

2. Measure and record the primary voltages **Vap**, **Vbp**, and **Vcp**.

node		Vap	Vbp	Vcp
Mag. V p-p				
Angle Deg.				

3. Measure and record the secondary voltages **Vas**, **Vbs**, and **Vcs**.

node		Vas	Vbs	Vcs
Mag. V p-p				
Angle Deg.				

Analysis, Part 1

Use the part 1 measurements to make the calculations below.

1. Calculate the primary currents, **Ia**, **Ib**, and **Ic**.

2. Calculate the secondary line to line voltages and the total power, P, delivered to the load.

3. Calculate the total complex power, **S**, and total average power, P. Calculate the circuit's power factor.

4. Calculate the efficiency of the circuit and of the 3-phase transformer.

5. Compare your results to a simulation.

Analysis, Part 2

Use the part 2 measurements to make the calculations below.

1. Calculate the primary currents, **Ia**, **Ib**, and **Ic**.

2. Calculate the secondary line to line voltages and the total power, P, delivered to the load. Calculate the efficiency of the circuit.

3. Calculate the total complex power, **S**, and total average power, P. Calculate the circuit's power factor.

4. Explain the voltage across the load resistor, Ra, when the primary winding of the phase **a** is open (R1 removed).

Analysis, Part 3

Use the part 3 measurements to make the calculations below.

1. Calculate the primary currents, **Ia**, **Ib**, and **Ic**.

2. Calculate the secondary line to line voltages and the total power, P, delivered to the load.

3. Calculate the total complex power, **S**, and total average power, P. Calculate the circuit's power factor.

4. Compare your results to a simulation.

LTspice Simulation

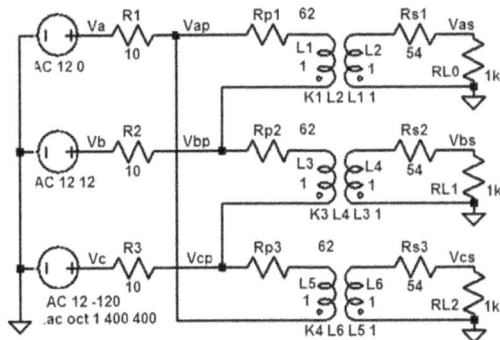

Normal

V(vap):	mag: 11.68	phase: 0.59°
V(vbp):	mag: 11.68	phase: 120.59°
V(vcp):	mag: 11.68	phase:-119.40°
V(vas):	mag: 18.12	phase: -28.07°
V(vbs):	mag: 18.12	phase: 91.92°
V(vcs):	mag: 18.12	phase:-148.07°

Open line (R1 open)

V(vap):	mag: 5.99	phase: 180°
V(vbp):	mag: 11.81	phase: 121.11°
V(vcp):	mag: 11.70	phase:-120.23°
V(vas):	mag: 9.06	phase: -88.07°
V(vbs):	mag: 18.12	phase: 91.93°
V(vcs):	mag: 9.06	phase: -88.07°

Experiment 5c: Open-Delta transformer

There is experiment uses two transformers to implement delta to delta configuration. This is called the "open delta" configuration.

Equipment and Parts

Function Generator, Oscilloscope, DMM, and Breadboard.
Resistors: Three 10Ω, three 1k, all ¼ watt, 5%.
Transformer, Two 500Ω CT to 500Ω CT, 400mW (see appendix 2).
    Recommended: ZICON 42TU500-RC (from Mouser Electronics)

Procedure

1. Connect the circuit in Figure 5-6 below. The amplitude of each phase is 12V p-p at the phase angles indicated.

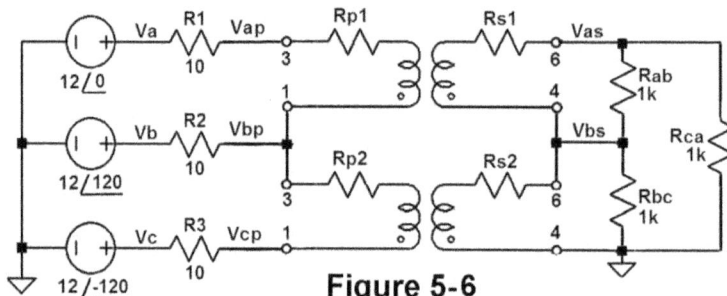

Figure 5-6

2. Connect oscilloscope channel 1 to **Va**. Trigger on channel 1. **Va** will be the reference phase for the entire experiment.

3. Measure and record the magnitude of **Va** and the magnitude and phase angle of **Vb** and **Vc**.

    Va:_____ V p-p **θa:** 0⁰   **Vb:**_____ V p-p **θb:**_____⁰

    Vc:_____ V p-p **θc:**_____⁰

5. Measure and record the primary voltages **Vap**, **Vbp**, and **Vcp**.

node	Vap	Vbp	
Mag. V p-p			
Angle Deg.			

6. Measure and record the secondary voltages **Vas**, and **Vbs**.

node	Vas	Vbs	Vcs
Mag. V p-p			
Angle Deg.			

Analysis

1. Calculate the primary currents, **Ia**, **Ib**, and **Ic**.

2. Calculate the secondary line to line voltages and the total power, P, delivered to the load.

3. Calculate the total complex power, **S**, and total average power, P. Calculate the circuit's power factor.

4. Calculate the efficiency of the circuit and of the open delta 3-phase transformer.

5. Compare your results to a simulation.

LTspice Simulation

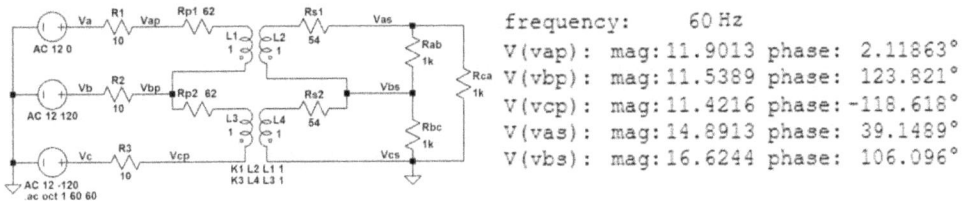

```
frequency:      60 Hz
V(vap):   mag: 11.9013 phase:  2.11863°
V(vbp):   mag: 11.5389 phase:  123.821°
V(vcp):   mag: 11.4216 phase: -118.618°
V(vas):   mag: 14.8913 phase:  39.1489°
V(vbs):   mag: 16.6244 phase:  106.096°
```

Vab = Vas − Vbs = 14.89∠39.15 − 16.62∠106.1

Vbc = Vbs = 16.62∠106.1

Vca = 0 - Vas = 0 - 14.89∠39.15 = 14.59∠-140.85

Appendix

Appendix 1: Three-Phase Sources

The Three-phase experiments in this book can done with a simple one-phase to Three-phase converter circuit described below. It may be built on a solder- less breadboard or purchased in kit or assembled form. Refer to the block diagram in figure 6-1 below.

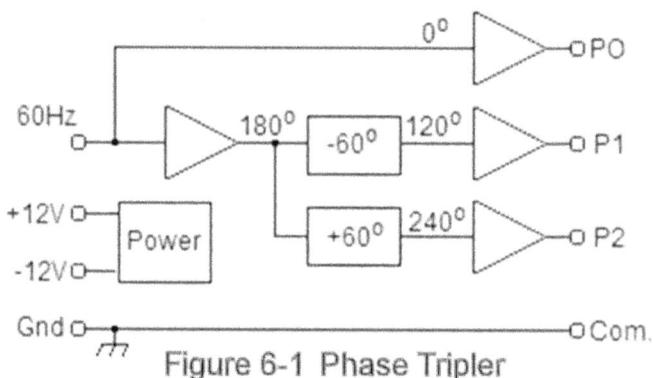

Figure 6-1 Phase Tripler

This phase tripler circuit converts single-phase AC to three-phase AC. The single-phase source may be a function generator or the ac voltage from a step-down transformer.   A 120VAC to 6.3VAC step-down transformer may be used with a voltage divider and a simple low-pass filter. See figure 6-3 on the next page.

The single-phase input is buffered by op-amp, U1A, and output at P0. Phase inverter, U1B, shifts the phase of the input by 180 degrees. This phase shifted voltage is applied to a 60 degree lag network to obtain a net phase shift of 120 degrees, and to a 60 degree lead network to obtain a net phase shift of 240 (-120) degrees.

The component values for the phase shift networks are chosen for the desired operating frequency. Part values are given for 60Hz or 400Hz operation.

The *Phase tripler* Three-phase source requires a ±12VDC, 200mA, power supply, and a single-phase 60Hz or 400Hz source.   If the output amplitude is less than 12V p-p, a ±9VDC supply can be used.

60Hz Parts List

U1, U2: L272M	R5, R8: 39K, ¼ watt, 5%.	C1, C2: 56nF, 2%.
R1, R2: 10K, ¼, 5%.	R6, R9: 47K, ¼ watt, 5%.	C3, C4: 100µF, 25V.
R3: 4.7K, ¼ watt, 5%.	R7: 27.4K: ¼ watt, 1%.	C5, C6, C7: 220nF, 10%
R4, 82.0K, ¼ watt, 1%.	R11, R12: 20K trim-pots.	D1, D2: 1N4001 diodes.

400Hz Modification: R4 = 18.8K, R7 = 5.9K, C1 and C2 = 39nF.

Resistor color code below. Check with the ohmmeter if uncertain.

1 Brown	6 Blue
2 Red	7 Violet
3 Orange	8 Grey
4 Yellow	9 White
5 Green	

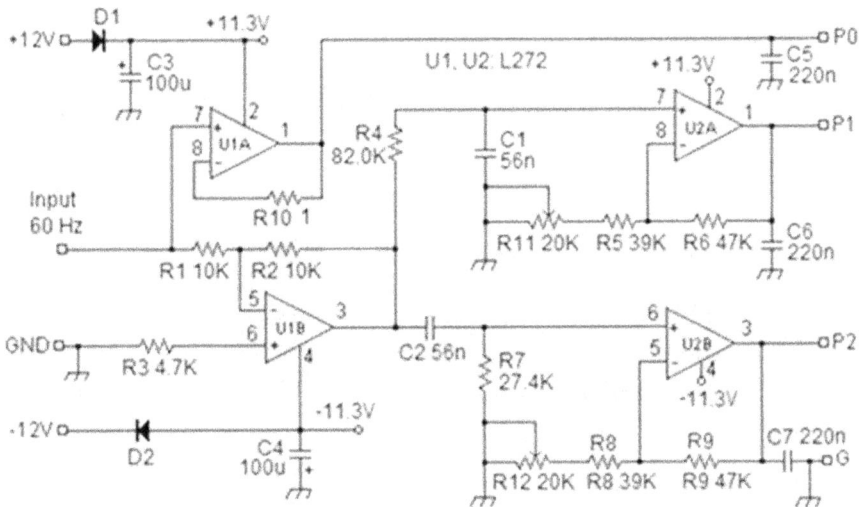

Figure 6-2

Transformer Option: A 6VAC wall transformer is recommended.

Capacitors are ceramic type, 25V minimum, and 10% tolerance. Resistors are ¼Watt, 5%.

Figure 6-3

60Hz Test and Calibrate

1. Connect the Phase Tripler to the power supply.   Connect a function generator (or transformer circuit) to the input and set it to produce a 12V p-p, 60Hz, sine wave. Connect the oscilloscope channel 1 to output P0 and channel 2 to P1.   Set both channels to AC input and 2 volts per division. S et the trigger to channel 1 and the time base to 2mS per division. Center both traces.   The output P0 should be exactly 12V p-p.

    Refer to the figures 6-4a and 6-4b below for steps 2 and 3. The zero crossings for P0, P1, and P2 are circled. In figure 6-4a P1 leads P0 and in figure 6-4b P2 lags P0.

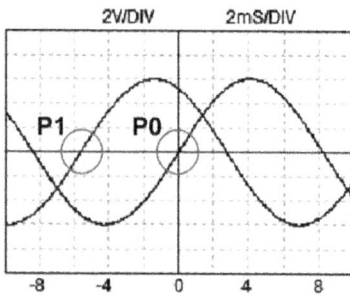 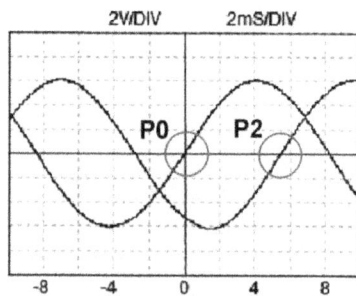

Figure 6-4a   Figure 6-4b

2. Adjust pot R11 so that the amplitude of output P1 is exactly 12V p-p. Check that the positive slope zero crossing of P1 occurs about 5.56mS before the positive slope zero crossing of P0. Refer to the diagram below.

3. Connect oscilloscope channel 2 to output P2. Adjust pot R12 so that the amplitude of output P2 is exactly 12V p-p. Check that the positive slope zero crossing of P2 occurs about 5.56mS after the positive slope zero crossing of P0.

400Hz Test and Calibrate

Same procedure as above applies except that the zero crossing times are 0.833mS instead of 5.56mS.

Specifications

Output voltage: 0 to 18V p-p, each phase output, P0, P1, P2 to G.
Output power: 125mW maximum, each phase output, P1, P2, P3.
Output frequency: 60Hz, ±1%.   P0 = 0^0. P2 = 120^0. P3 = 240^0.
*Output frequency: 400Hz, ±1%.   P0 = 0^0. P2 = 120^0. P3 = 240^0.
Input: 0 to 18V p-p, 60Hz, ±1%.
*Input: 0 to 18V p-p, 400Hz, ±1%.
Input power: +12VDC and -12VDC, 100mA maximum
*(400Hz option).

Load Limitations

Loads may be any configuration with the limitation that the power should not exceed 125mW per phase.   At 12 volts peak-to-peak this is about 30mA RMS per phase.   Each phase output is supplied by a very low output impedance power op-amp. Excessive current will cause the op-amp to over heat and shut down.   The op-amp may resume operation when the excessive current is removed and the op-amp cools down.

The lowest impedance per phase that can be connected to the power supply depends on the load configuration. For example, at 12V p-p, the lowest impedance for a four-wire wye would be about 140 ohms per phase (4.24VRMS/0.03A RMS). As a general rule, any impedance greater than 300 ohms per phase is safe to use with any load configuration. ***The op-amps will get hot and should not be touched with loads above 100mW per phase.***

Another consideration is the power dissipation of the load components. Lab experiments using ¼ watt resistors should be designed so that the resistors don't get too hot.   To avoid hot op-amps and hot ¼ watt resistors, use impedances greater than 400 ohms per phase.

Measurements

This is basically a wye-connected source with a common connection that is also the circuit ground for the power supply and op-amps. Non-isolated instrument grounds cannot be connected to the output of an op-amp because that would cause a short circuit.

<u>Measurements should be made with respect to the common ground, G, unless the measuring instrument is known to be isolated.</u>

The Phase Tripler may be used as a delta source by not using the neutral, G, connection. However, measurements should be made with respect to the neutral, G, unless the instrument is isolated from the lab electrical system ground.

Example:

A delta load (Ra, Rb, Rc) is connected to the Phase Tripler as shown in figure 6-5 on the right. Resistors Rw represent line resistance. If the measuring instrument is not known to be isolated, all measurements must be made with respect to the common (G). Line to line voltages can be determined by phasor calculations:

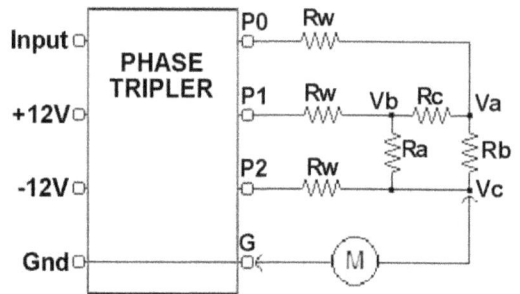

Figure 6-5

$$V_{ab} = V_a - V_b, \qquad V_{bc} = V_b - V_c, \qquad V_{ca} = V_c - Va.$$

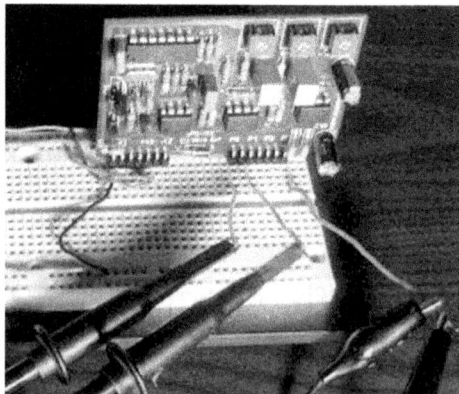

ZAP Studio Three-Phase Sources

Phase Tripler III BD-3A60

The Phase Tripler is easy to build and easy to use. Its circuit board is silk screened with part numbers. It can be used as a voltage source for Three-phase experiments and demonstrations.

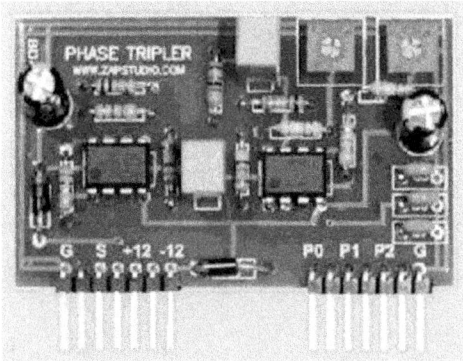

Board size: 2.5 X 1.5 in.
DC input: +12V and -12V
AC input: 0 to 18V p-p
60Hz, 1-phase
Output: 0 to 18V p-p
60Hz, 3-phase
125mW per phase

Also available for 400Hz:
model BD-3A400.

Three-Phase Voltage Source 3PS-02A60

This source is similar to the Phase Tripler and it has the same pin-out as the Phase Tripler. It has an on-board synthesized 60Hz source so that an external 60Hz input is not required. Also available as a kit.

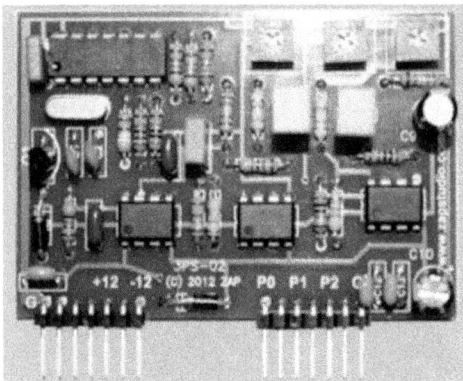

3PS-02A60
Board size: 2.8 X 1.9 in.
DC input: +12V and -12V
Output: 12V p-p, 3-phase
250mW per phase
Frequency error: <0.5%
Phase error: <2%

Also available for 400Hz:
Model 3PS-02A400.

The 400 Hz version is ideal for experiments and demonstrations which use small transformers (200mW).

These sources are available at: www.zapstudio.com

Appendix 2: Parts Information

The part values chosen for the experiments in this book are based on the power output capability of the three-phase sources used and that the parts don't get too hot to touch.

The *Phase Tripler* outputs up to 125mW per phase which is about 30mA RMS per phase at 12V p-p (4.24V RMS). Minimum load impedances are given for an output of 12V p-p, phase to neutral, in the table below.

Connection	Load	V p-p	V rms	I rms/Phase	P/Phase
Line to Neutral	142	12V	4.24V	30mA	125mW
Line to Line	432	20.8V	7.35V	17mA	125mW

Transformers

Mouser Stock No.	Impedance		Resistance	
	Primary	Secondary	Primary	Secondary
42TU500-RC	500Ω	500Ω	35Ω	28Ω

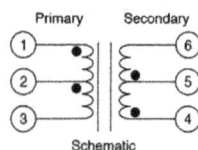

- Core type / size: EI-24
- Impedance variation: ±10% @ 1KHz
- D.C. resistance: ±15%
- Max. output: 460mW
- Frequency response: ±3dB, 300-3.4KHz @ 1KHz 0dB

Mouser Stock No.	Impedance (Ω)		Resistance (Ω)		Turns
	Pri.	Sec.	Pri.	Sec.	Pri.:Sec.
42TM016-RC	600	600	65	55	1:1

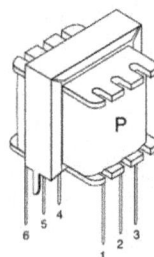

- Core type/size: EI-19
- Impedance variation: ±10% @ 1KHz
- DC resistance tolerance: ±15%
- Frequency response: ±3dB, 300Hz~3.4KHz @ 1KHz 0dB
- Maximum output: 200mW

The recommended transformers below are compact and have wire leads which can be directly plugged into a breadboard. They both work well at 400Hz. The 42TU500-RC may be used at 60Hz but with lower efficiency.

Inductors

The magnitude of an inductor's impedance is given by: $|Z_L| = \sqrt{R_w^2 + X_L^2}$.

For convenience, the magnitude of the impedance at 60Hz and 400Hz of a selection of inductors is given in the table below. These are manufactured by Fastron: www.fastrongroup.com.

These inductors are available from Mouser Electronics at a unit price of $1.38 per unit (January 2014 price).

No	Mouser Part Number	L mH	R_W Ω	60Hz X_L	60Hz $|Z_L|$	400Hz X_L	400Hz $|Z_L|$
1	434-02-223J	22	20	8.3	21.7	55.3	58.8
2	434-02-333J	33	26	12.4	28.8	82.9	86.9
3	434-02-563J	56	58	21.1	61.7	140.7	152.2
4	434-02-683J	68	66	25,6	70.8	170.9	183.2
5	434-02-823J	82	71	30.9	77.4	206.1	218
6		100	120	37.7	125.8	251.3	278.5

The table below provides the manufacturer data for the inductors above.

No	Part No	Inductance L (mH)	f_L (MHz)	Tol ± (%)	Q min	f_Q (MHz)	DCR max (Ω)	Rated DC Current (mA)
1	07M-223K-50	22	0.02	10	100	0.079	19.5	21
2	07M-333K-50	33	0.02	10	100	0.079	26.0	17
3	07M-563K-50	56	0.02	10	100	0.079	58.0	12
4	07M-683K-50	68	0.02	10	70	0.079	66.0	12
5	07M-823K-50	82	0.02	10	70	0.079	71.0	10
6	07M-104K-50	100	0.02	10	55	0.030	120	7

Capacitors

A 1μF capacitor has a reactance of 2653Ω at 60Hz and 398Ω at 400Hz. Capacitors in the 1μF to 10μF are needed for practical experiments at 60Hz. These must be non-polarized and preferably have a tolerance of 10% or better. The tolerance is not critical if an accurate capacitance meter is available.

A section of the Mouser on-line catalog (January 2014) below shows a sampling of capacitors available. TDK MLCC Capacitors.

MOUSER STOCK NO.		Value	Volt.	Tol.	Price Each	
Mfr.	Mfr. Part No.	(pF)	(Vdc)	(%)	1	50
810—	FK11X7R2A105K	1µF	100	10%	.49	.374
810—	FK11X7R1H225K	2.2µF	50	10%	.61	.468
810—	FK11X7R1H335K	3.3µF	50	10%	.61	.468
810—	FK11X7R1H475K	4.7µF	50	10%	.73	.56
810—	FK11X7R1E685K	6.8µF	25	10%	.88	.654
810—	FK11X7R1C106K	10µF	16	10%	.62	.457

On-Line Part Sources

All Electronics

Allied Electronics

Digi-Key Electronics

Jameco Electronics

Mouser Electronics

Newark Electronics

Radio Shack

Appendix 3: Phasor Algebra

Polar to Rectangular Conversion

$|A| \angle \theta = |A| \cos \theta + j|A| \sin \theta$

Example: $10 \angle 30^0 = 10 \cos 30 + j10 \sin 30 = 8.66 + j5$

Rectangular to Polar Conversion

$A + jB = \sqrt{A^2 + B^2} \, \angle \arctan\left(\dfrac{B}{A}\right)$

Example: $8.66 - j5 = \sqrt{8.66^2 + 5^2} \, \angle \arctan\left(\dfrac{-5}{8.66}\right) = 10 \angle -30^0$

Addition and Subtraction

$(A + jB) + (C + jD) = (A + C) + j(B + D)$

$(A + jB) - (C + jD) = (A - C) + j(B - D)$

Examples:

$(5 + j8) + (4 - j6) = (5 + 4) + j(8 - 6) = 9 + j2$

$(5 + j8) - (4 - j6) = (5 - 4) + j(8 + 6) = 1 + j14$

Polar Multiplication

$(A \angle \theta)(B \angle \phi) = AB \angle (\theta + \phi)$

Examples:

$(9 \angle 30^0)(4 \angle 20^0) = 36 \angle 50^0, \qquad (7 \angle 45^0)(5 \angle -20^0) = 35 \angle 25^0$

Polar Division

$\dfrac{(A \angle \theta)}{(B \angle \phi)} = \dfrac{A}{B} \angle (\theta - \phi)$

Examples:

$\dfrac{(9 \angle 30^0)}{(3 \angle 45^0)} = 3 \angle -15^0, \qquad \dfrac{(8 \angle 45^0)}{(4 \angle -20^0)} = 2 \angle 65^0$

Rectangular Multiplication

$$(A+jB)(C+jD)=(AC-BD)+j(AD+BC)$$

Examples:

$$(5+j8)(4-j6)=(20+48)+j(-30+32)=68+j2$$
$$(2-j4)(4-j6)=(8-24)+j(-12-16)=-16-j28$$

Rectangular Division

$$\frac{(A+jB)}{(C+jD)}=\frac{(A+jB)(C-jD)}{(C+jD)(C-jD)}=\frac{(A+jB)(C-jD)}{C^2+D^2}$$

Example:

$$\frac{(5+j8)}{(4-j6)}=\frac{(5+j8)(4+j6)}{(4-j6)(4+j6)}=\frac{(20-48)+j(30+32)}{4^2+6^2}=\frac{-28+j62}{52}=-0.5385+j1.1923$$

Appendix 4: Measuring Part Values

In circuit analysis we often assume that we have ideal components. An ideal resistor has only the property of resistance, which is independent of frequency. In reality, a resistor also has the properties of inductance and capacitance, but these properties are not significant at low frequencies. Experiments in this book are done at frequencies below 10KHz so that the values and properties of the components will be relatively independent of frequency.

A good quality digital multi-meter can read resistor values with an accuracy of better than 1%. The accuracy of the meter can be tested with a set of 1% resistors.

We want to know that the circuit performance will be as predicted by the theoretical analysis. In order to compare the measured results to theory, we need to know the values and properties of the components used.

Bridge type instruments are available for inductance and capacitance measurement, but they are expensive. An alternative to these instruments would be to use available equipment, such as a signal generator and an accurate oscilloscope, to make the measurements.

Capacitance Measurement

The circuit and phasor diagrams used for determining the capacitor's value are shown on the below. The current, I, is the same in all of the components of the series circuit, so $V_C = I X_C$, and $V_R = I R$.

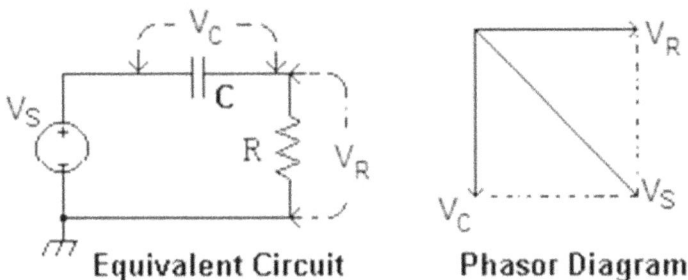

Equivalent Circuit   Phasor Diagram

In the diagram above, the frequency, f_C, is set to where $|V_C| = |V_R|$. At that frequency $|V_X| = 0.707|V_S|$ and $X_C = R$.

R can be measured with an ohmmeter and C can be calculated:

$$X_C = R = \frac{1}{2\pi f_c C} \text{ , therefore } C = \frac{1}{2\pi f_c R} \text{ .}$$

Inductance Measurement

At low frequencies an inductor may be approximated by an inductance in series with a resistance. The resistance is mainly due to the wire winding of the inductor. This resistance does increase with frequency, however it is approximately constant for the low frequencies used in the experiments in this book.

The diagrams below shows the equivalent circuit and phasor diagram used for determining the inductance of the inductor. Note that we can't directly measure the voltage across R_W (inductor wire resistance), but we can measure the voltage across R_X. The inductor's reactance, X_W, is equal to ($R_W + R_X$) when $|V_X + V_W| = 0.707|V_S|$. The required value of $|V_X|$ can be calculated.

$$|V_X| = \left(\frac{R_X}{R_X + R_W}\right)\frac{|V_S|}{\sqrt{2}}$$

Equivalent Circuit   Phasor Diagram

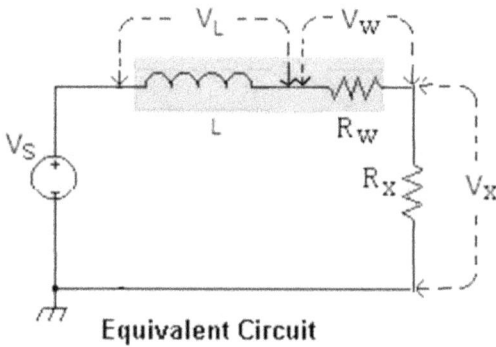

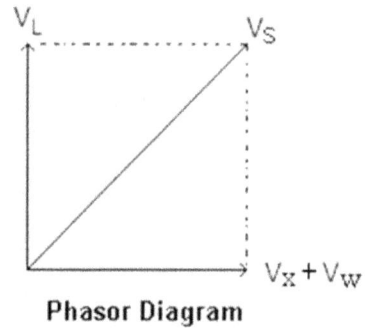

We can then tune the frequency of the function generator so that $|V_X|$ is equal to the required value. We can use that frequency, f_c, and the measured values of R_W and R_X to calculate L.

$$X_L = 2\pi f_c L = (R_W + R_X), \qquad L = \frac{(R_W + R_X)}{2\pi f_c} \text{ .}$$

110

An experiment to measure the value of an inductor or capacitor works best if it is done near the operating frequency of the circuit in which the components will be used. Also the approximate value of the inductance or capacitance should be known.

Example Capacitance Measurement

A capacitor with a known value of 0.1µF will be measured. This method is useful to get a more accurate measured value of a capacitor whose approximate value is known.

Equipment and Parts

Function Generator, Oscilloscope with 10X probes, Breadboard.
C = 0.1 µF, 5%, capacitor, polyester film. R = 1000 ohms, 5%, ¼

Procedure

1. Measure the value of the 1000 Ω resistor, Rx, and record below.

   Rx _____

2. Connect the   circuit on the right.
   Set the generator to   produce a
   1600Hz, 5.00 V p-p   sine wave
   with no offset.

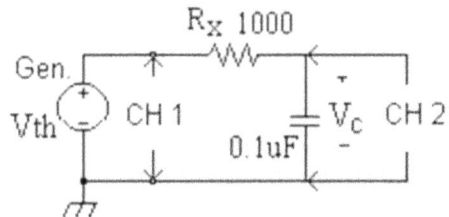

3. Carefully vary the function generator frequency, f_C, while observing the amplitude of the voltage, V_C, on channel 2 of the oscilloscope. Adjust the frequency so that $|V_C|$ is exactly 3.54 V p-p.

4. Check that the channel 1 voltage is 5 V p-p. If not, reset it to 5V p-p and re-adjust the frequency to get   3.54-volts p-p on channel 2.

5. At this frequency, f_C, the magnitude of the capacitive reactance equals the value of the resistance, R_X. Check that the phase angle θ_X is close to 45 degrees.

   Calculate and record the capacitance:

   $$C = \frac{1}{2\pi f_C R_x}$$ C = _____ θ_C = _____

Note the display below. The oscilloscope's "measure" feature was used to display the frequency and amplitudes on the right side of the screen *(Refer to your oscilloscope's user manual)*. Note the phase shift between channel 1 and channel 2. Channel 2 leads channel 1 by 45 degrees.

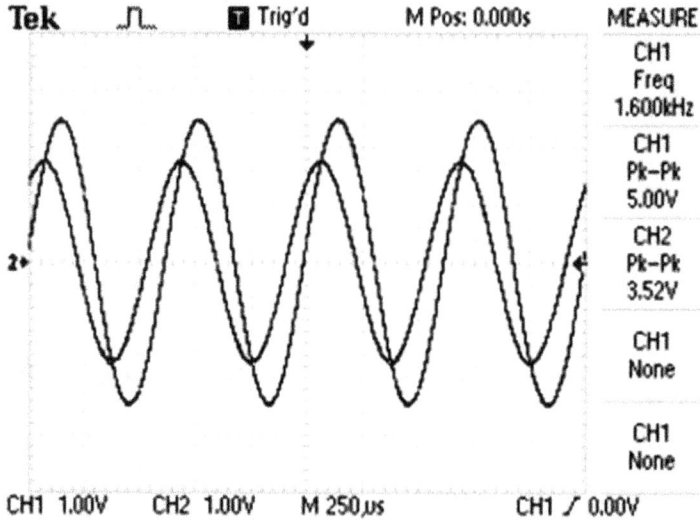

The results of step 4 should be accurate to about 3% when using   an accurate oscilloscope.

Example Inductance Measurement

A 100mH, 10% tolerance inductor is measured. For best results the measurement frequency should be approximately the same as the frequency of the inductor application.

Equipment and Parts

Function Generator, Oscilloscope with 10X probes, Breadboard.
L = 100 mH, 30mA, Mouser Xicon 43LJ410 or equivalent.

Procedure

1. Use the same 1000 Ω resistor, R_X, which you used in part 1. Measure the resistance, R_W, of the inductor.

   R_W: _____

2. Connect the circuit on the right. Connect channel 1 and channel 2 of the oscilloscope as shown. Set the function generator to produce a 1600Hz, 5V p-p sine wave.

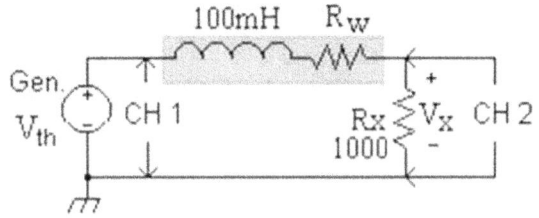

Calculate and record the magnitude of V_X: $|\mathbf{V_x}| = \dfrac{R_X}{R_X + R_W} \cdot 3.54$.

$|\mathbf{V_x}| = $ _____

3. Vary the function generator frequency while observing the amplitude of the voltage, V_X, on channel 2 of the oscilloscope. Adjust the frequency so that $|\mathbf{V_x}|$ is exactly equal to the voltage calculated above (channel 1 voltage must be exactly 5 V peak-to-peak).

At this frequency, f_C, the inductive reactance equals the resistance of $(R_X + R_W)$. Check that the phase angle θ_X is close to 45 degrees. Calculate the inductance (use the measured values of R_X and R_W):

$$L = \frac{(R_X + R_W)}{2\pi f_C}.$$ $L = $ _____ $\theta_X = $ _____

The results of step 3 should be accurate to about 3% when using an accurate oscilloscope.

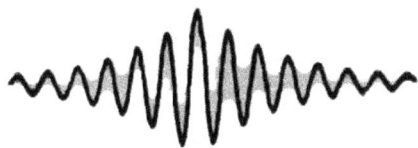

www.ingramcontent.com/pod-product-compliance
Lightning Source LLC
Chambersburg PA
CBHW051222200326
41519CB00025B/7219